Consulting Authors

William C. Kyle, Jr.
E. Desmond Lee Family Professor of Science Education
University of Missouri—St. Louis
St. Louis, Missouri

Brenda Webb, Assistant Professor
Kilby Laboratory School
University of North Alabama
Florence, Alabama

The McGraw-Hill Companies

Program Reviewers

Kimberly Bailey
Teacher
Ariel Community Academy
Chicago, Illinois

Suzanne Giddens
Teacher
Jenks West Elementary
Jenks, Oklahoma

David Maresh
Teacher
Yucca Valley Elementary
Yucca Valley, California

M. Kate Thiry
Teacher
Wright Elementary School
Dublin, Ohio

April Causey Trull
Teacher
Bolivia Elementary School
Bolivia, North Carolina

Constance W. Zehner, M.S.
CZ Consulting
Houston, Texas

Assessment Specialist

Michael Milone, Ph.D.
Placitas, New Mexico

www.sra4kids.com

Send all inquiries to:
SRA/McGraw-Hill
8787 Orion Place
Columbus, OH 43240-4027

Printed in the United States of America.

ISBN 0-07-572920-2

1 2 3 4 5 6 7 8 9 RRC 08 07 06 05 04 03 02

The McGraw·Hill Companies

TABLE OF CONTENTS

Investigating Skills

TABLE OF CONTENTS

Reading and Thinking Skills

Writing and Research Skills

Chart and Graph Skills

Test-Taking Strategies

Investigating Skills

Skill 1

HOW TO

Use a Model

When Fronts Collide

Suppose the weather has been warm and sunny, and you and your family are planning to spend the day at the park. You're on your way out the door when you catch sight of a weather person on TV pointing to a map of the United States and saying, "A cold front rapidly moving in from the northwest will meet the mass of warm air that now covers our area. A line of thunderstorms will develop early this afternoon where the two fronts collide." Will this affect the weather where you are? You'd better watch the rest of the forecast to find out.

A **meteorologist** is a person who studies and predicts the weather. Meteorologists use different kinds of maps to help them explain weather conditions. These weather maps are models that show what the weather will be like across the country and in your local area. A **model** is a structure or a picture that shows an object, event, or idea. Models can help explain concepts that are hard to see or difficult to understand. Weather models make it easy to see weather patterns that are developing across the country.

Meteorologist

Because it is impossible to see all the weather conditions across a country or even one state, weather maps are used to show images that represent weather conditions. Your local newspaper or television station might have daily weather maps that show you what you can expect in your area. Most people find it easier to understand models, maps, photographs, and drawings of weather conditions than written or spoken descriptions.

Look at the weather map below. Think about how you would describe in words what is shown on the map.

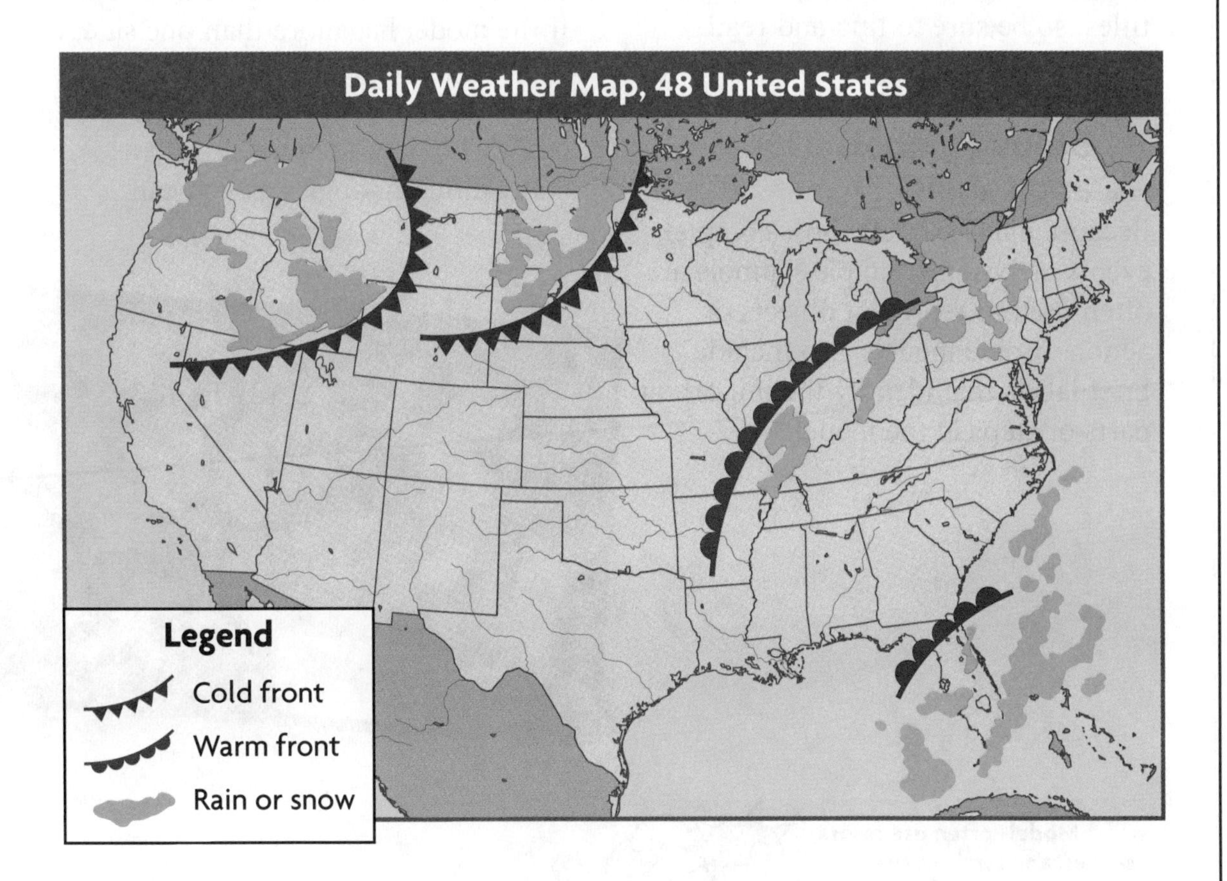

STEPS IN Using a Model

Follow these steps to learn how to use a model.

1 Examine the Model

Study the model to get a general sense of what the model is showing. A model is almost always much simpler than the event or system it represents. Some models will have captions or titles, so be sure to find and read these.

2 Study Symbols and Labels

Because many models show complex events or systems, simple symbols are often used to represent objects or actions. Some models also include brief labels that identify the important parts or steps in the model.

3 View All Sides

Look at every part of the model to fully understand what is being shown. If the model has more than one side, or if the model can be picked up and turned around, turn or move the model. Look at the model from as many different angles as you can.

TIP Models often use colors, shapes, and symbols that people understand without the need for an explanation. For example, red often represents heat, whereas blue often represents cold.

Models such as weather maps use pictures and symbols to give information. Read below to see how a student in Iowa put into words what she saw on a weather map of her state.

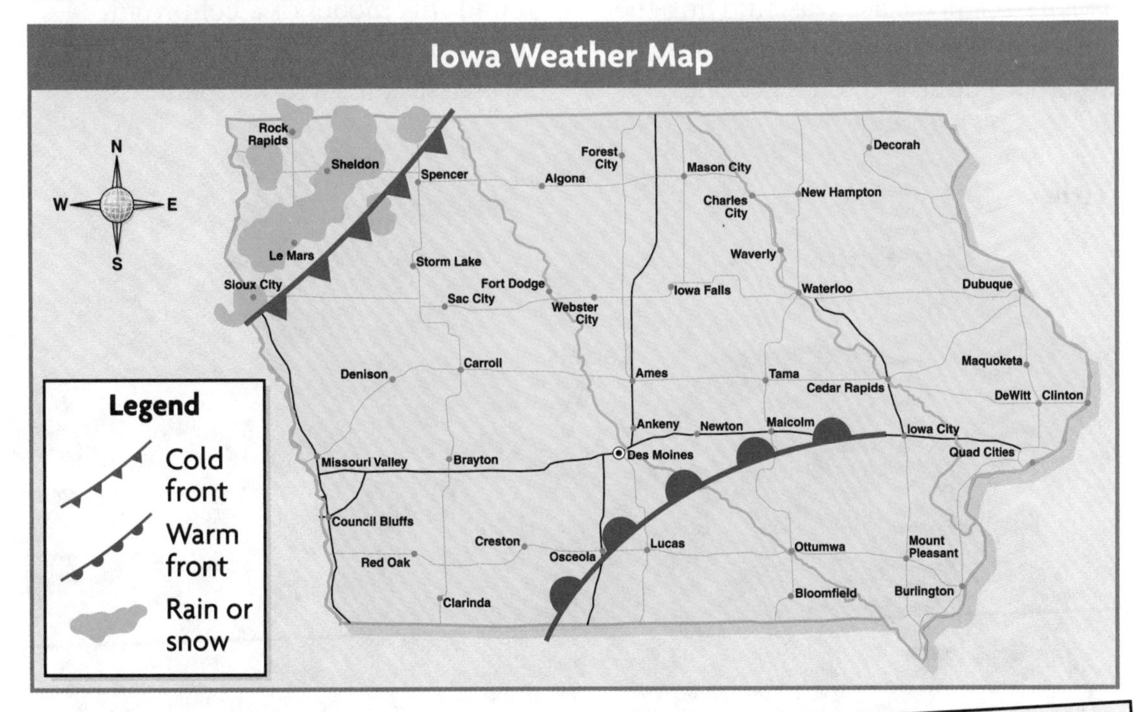

The blue line shows a cold front, and the blue triangles on the line show the direction in which the cold front is moving. The curved red line shows a warm front, and the red bumps on the line show the direction in which the warm front is moving. The green shading shows areas of rain. The weather map is showing that a cold front and rain are moving in from the northwest to meet the warm front that is now in the center of the state. This means that there probably will be thunderstorms in Iowa.

EXAMPLE OF **Using a Model**

Read the following passage to see how one student used a model to understand how a cold front develops.

Erica's science class was studying the different types of fronts that cause weather patterns to change. She found this model of a cold front, which helped her better understand air masses and weather patterns.

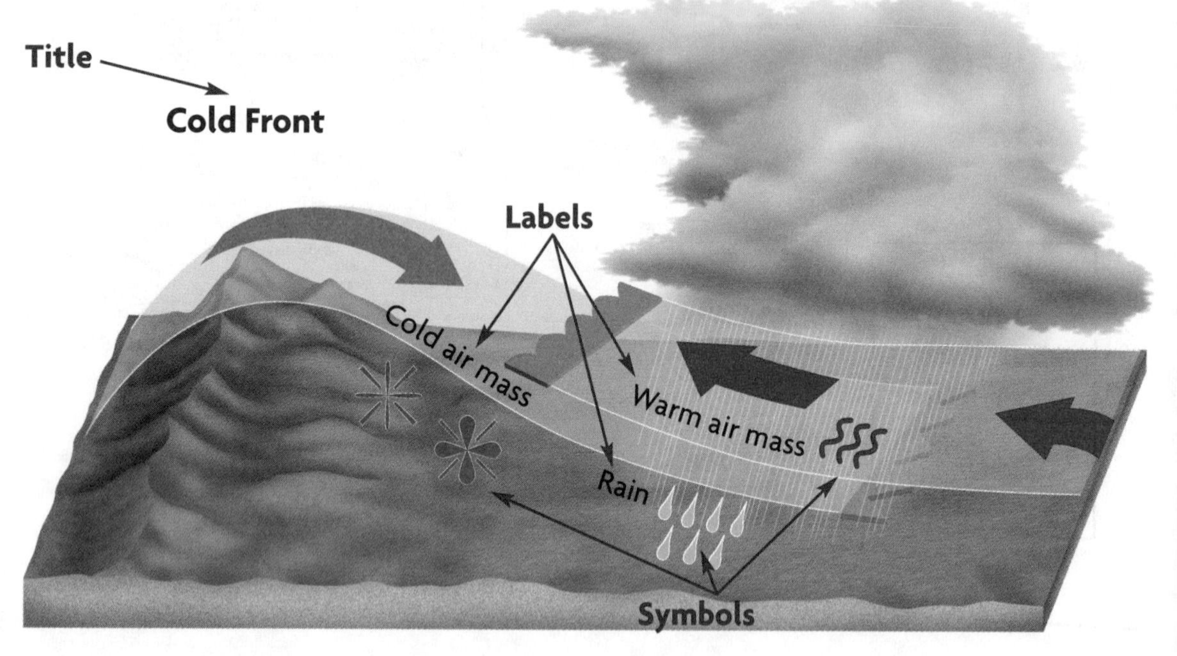

Once she had studied the model, Erica could picture the formation of a cold front. She was able to use the model to write the following definition for homework:

A cold front forms when a cold air mass pushes into a warm air mass. The warm air mass rises over the cold air mass because it is less dense. As the warm, damp air mass rises, thick clouds form. These clouds produce heavy rain.

USE THIS SKILL

Use a Model

Examine the model of temperatures expected in the continental United States on one day in October. Then answer the questions below.

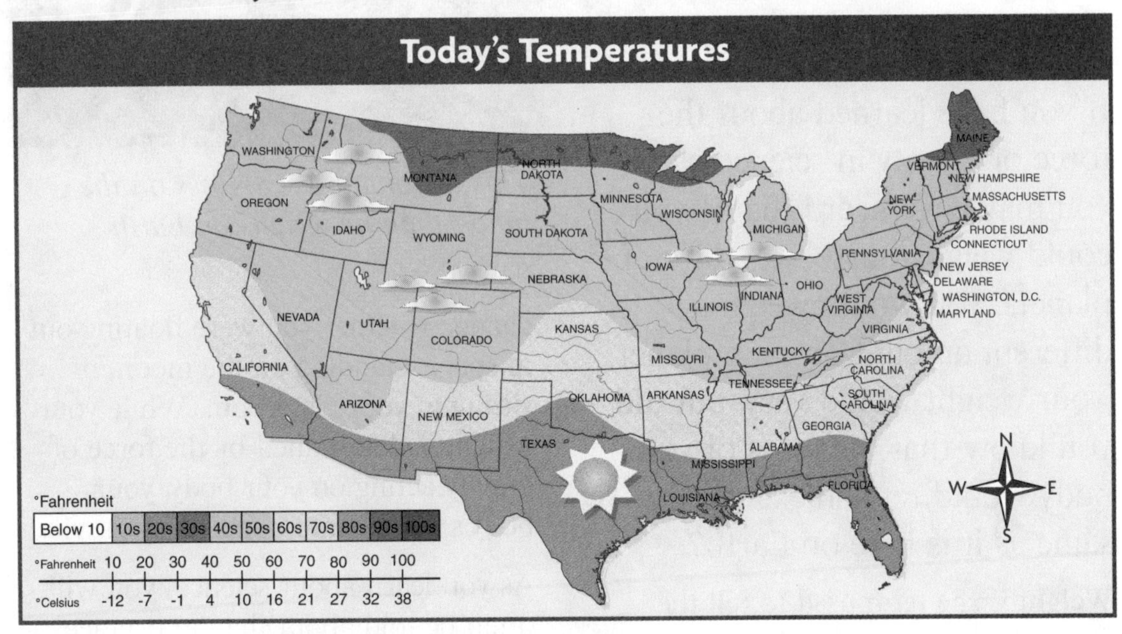

Answer these questions using the map above.

1. What temperature range is shown in green?
2. What temperature range is shown in yellow?
3. Name three states with temperatures in the 30s.
4. What is the temperature expected in Florida?
5. What is the temperature expected in Ohio?
6. Which state has warmer temperatures, New York or Virginia?

TEST TIP

Some tests may ask you to use a model to answer questions. Quickly look over the model before you begin answering the questions. Look back at the model carefully as you answer the questions.

Skill 2

HOW TO Measure

Balancing Mass

If you have learned about the force of gravity in science class, you may have heard that if you could weigh yourself on different planets, you would weigh a different amount on each planet. Your weight would vary, but did you know that the mass of your body would remain exactly the same as it is here on Earth?

Weight is the term used to tell the force of gravity on an object. A 10-kilogram metal ball has a stronger pull of gravity acting upon it than a 2-kilogram plastic ball does. The two balls may be the same size, but we can pick them up and feel the difference in their weights. The heavier ball has a greater force of attraction (gravity) to our planet than the lighter ball.

The **mass** of an object tells the amount of matter in that object. The amount of matter in your body would not

This astronaut weighs less on the moon than he weighs on Earth.

change whether you were floating out in space, standing on the moon, or sitting in your classroom. While your weight is determined by the force of gravity acting on your body, your body's mass is not affected by gravity.

As you learn about science, you will often be told about the importance of taking accurate measurements when studying matter and doing experiments. Weight and mass are both properties of matter that can be measured, although they should be measured using different kinds of scales. The scales at a doctor's office, in grocery stores, bathroom scales, and hanging "spring" scales measure weight. A balance scale allows you to compare the mass of an object to be measured to the known mass of another object.

STEPS IN **Measuring**

Follow these steps to measure the mass of a solid substance using a balance scale.

1 Zero the Balance

Make sure that the two pans on the balance are empty. The pointer must be aligned with the middle mark on the scale.

2 Add the Object

Place the substance to be measured in one of the pans on the balance scale. Place a standard mass in the other pan. If the pointer is to the left of the middle mark, add more standard masses. If the pointer goes right of the middle mark, take masses away. Keep adding standard masses to this pan until the pointer is lined up with the middle mark on the scale. This is how you can tell that the two pans are balanced.

3 Record Data

Add up the amounts of the standard masses in the pan. This number is the mass of the solid substance. Record this number. Make sure to use the correct units of measurement.

4 Clear the Balance

Before you make the next measurement, you must clear and zero the balance scale. Remove your measured substance and all the standard masses from the balance pans. Make sure the pointer again aligns with the middle mark. You are now ready to measure another object.

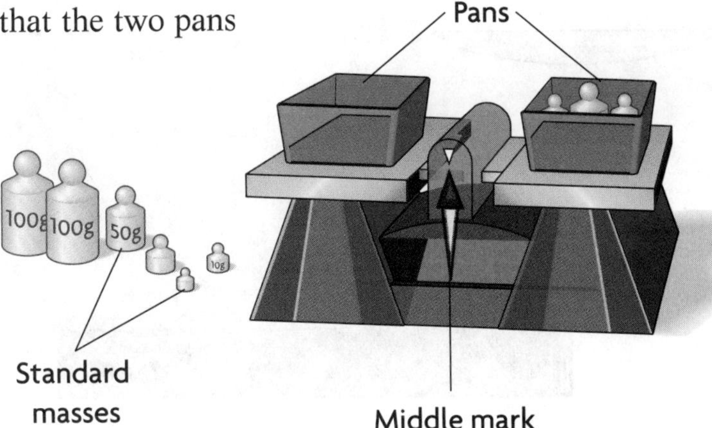

Pan balance

EXAMPLE OF Measuring

See how one student used a balance scale to measure the mass of a solid substance.

Zero the balance → *While doing a science experiment, Adam needed to measure the mass of some salt crystals using a pan balance. First Adam made sure that the pans were empty and the pointer on the balance aligned with the middle mark.*

Add the object → *Adam then placed the mound of salt on one pan of the balance. On the other pan, he placed one 100 gram standard mass. The pointer moved toward the middle mark but was still to the left of it. He added another 100 gram mass, then added a 10 gram mass. The pointer now aligned with the middle mark. Adam counted the number of grams of the standard masses and got a total of 210. He recorded the mass of*

Record data → *the mound of salt as 210 grams. Adam then removed the masses and the salt from the pans and checked to see that the pointer was lined*

Clear the balance → *up exactly with the middle mark.*

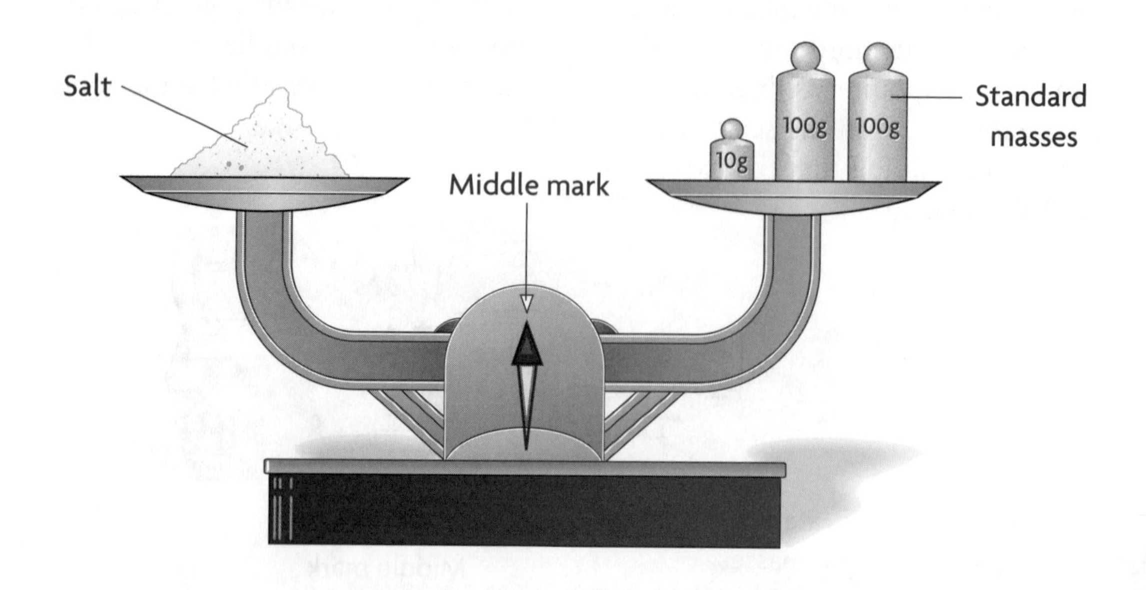

USE THIS SKILL

Measure

Look at the drawings and answer the questions below.

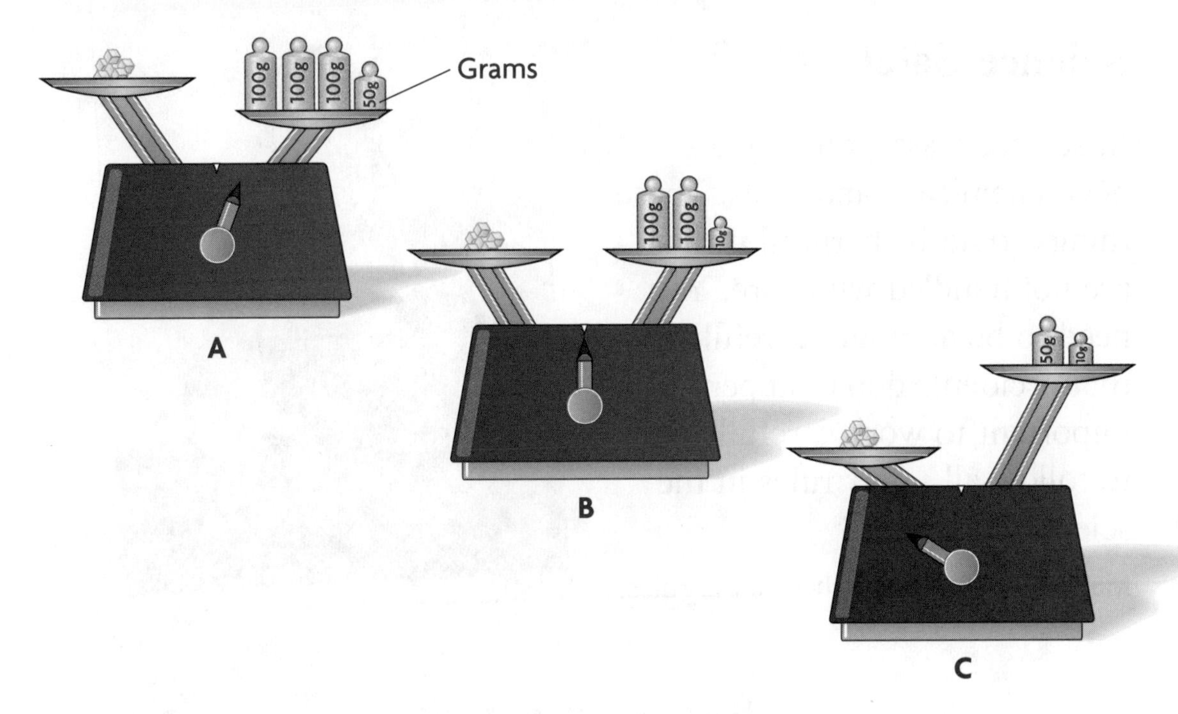

1. What is the mass of the crystal?
2. Which drawing, A, B, or C, shows too many masses on the balance?
3. Which drawing, A, B, or C, shows too few masses on the balance?
4. What would the mass of the crystal be on the moon?

TEST TIP	**Some tests may ask you to measure something or to read measurements. When answering these questions, make sure that you use the correct units of measurement.**

Skill 3

HOW TO

Work Safely

Science Safety

In science class, you may use heat, chemicals, and tools. These things could be harmful if they are not handled with care. You need to be alert and careful so that accidents don't happen. It is important to work carefully and to follow all safety rules in the science classroom.

Make sure you follow these basic rules:

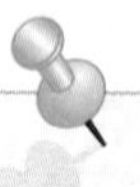

Basic Science Safety

1. Never use heat, chemicals, or other tools unless your teacher or another adult is present.
2. Follow all safety rules that are posted in the room.
3. Always use the protective equipment required.
4. Use materials only in the way that they are meant to be used in an experiment.
5. Tell your teacher right away if there is an accident or injury, even a small one.

STEPS IN Working Safely

Follow these steps to learn how to work safely in the science classroom.

1 Know How to Find Safety Equipment

Ask your teacher to show you the location of all safety equipment in the classroom. Learn how to use each item.

2 Read the Steps

Read all the steps for any experiment before you begin. Make sure you understand all instructions and safety symbols shown.

3 Prepare and Protect

Use all the protective gear that is recommended for the experiment. Before beginning the experiment, during the experiment, and while doing cleanup, make sure you do the following:

- ✓ Tie back long hair and loose clothing.

Clothing Protection Safety

- ✓ Wear safety goggles, a lab apron, and insulated gloves.

Eye Safety, Skin Protection Safety

- ✓ Keep all materials away from flames and heat sources.

Fire Safety

- ✓ Use tongs or a pot holder to pick up hot items.

Thermal Safety

- ✓ Always slant test tubes away from yourself and others.

Thermal Safety, Fume Safety

- ✓ Apply cold, running water to any minor skin burn.

- ✓ Never inhale chemicals or put them close to your nose or eyes.

Chemical Safety, Fume Safety, Poison Safety

- ✓ Tell your teacher immediately if you spill a chemical on your skin or clothing.

Chemical Safety

- ✓ Never taste any materials used in an experiment.

Poison Safety

- ✓ Return all chemicals to your teacher at the end of an experiment.

Chemical Safety

EXAMPLE OF Working Safely

See how one student followed safety procedures while doing a science experiment.

Colin was planning to conduct a simple activity to find the effect of sulfuric acid on rock.

Try It!

Does acid cause rocks to weather rapidly?

1. Place 100 g of shale chips in a 250-mL beaker.
2. Add 100 mL of a dilute solution of sulfuric acid.
3. Observe the mixture for 2 to 3 minutes.
4. Stir the mixture and allow it to sit until it stops bubbling.
5. Pour the mixture through a piece of filter paper.
6. Rinse the shale chips with water and dry them.
7. Weigh the chips and compare their mass now with their mass before.

Prepare and protect — Before beginning the activity, Colin carefully read the entire procedure and noted all the safety symbols. He knew how to find and use all the safety equipment in his classroom. He gathered all the materials that he needed except the sulfuric acid, and then he put on an apron, safety goggles, and gloves. When he was ready to begin, Colin got the sulfuric acid from his teacher.

Eye safety

Clothing protection

Skin protection

Colin then began the activity. He kept the bottle of sulfuric acid tilted away from him while he was pouring the acid into the beaker. Throughout the activity, Colin was careful not to breathe in any of the acid's fumes.

Chemical safety

Fume safety

When Colin was finished with the activity, he returned the sulfuric acid to his teacher. He cleaned the beaker and his gloves and goggles with hot, soapy water before returning them to the shelf. He also put the shale chips back on the shelf and threw the filter paper in the trash.

USE THIS SKILL

Work Safely

For each experiment listed below, tell which of the following safety tools or procedures would be used.

Clothing Protection Safety

Fire Safety

Thermal Safety

Skin Protection Safety

Eye Safety

Fume Safety

Poison Safety

Chemical Safety

1. Mixing vinegar and baking soda to observe the chemical reaction
2. Boiling water on a hot plate to observe changing states of matter from liquid to gas
3. Mixing different kinds of natural inks and dyes to stain cloth or strips of paper
4. Using a Bunsen burner while testing the effects of heat on different kinds of metal shavings
5. Using litmus paper to test the acidity of an unknown chemical
6. Mixing oil, water, and food coloring while studying mixtures and solutions
7. Burning a piece of wood or magnesium metal while observing chemical reactions
8. Measuring or pouring ammonia, rubbing alcohol, or hydrogen peroxide

TEST TIP

If a test question asks you what kind of tools or equipment you would use for a specific science activity or experiment, think about what the question is asking. Choose the tool that would be most helpful to complete the task.

Skill 4

HOW TO

Make a Hypothesis

Life Comes Only from Other Life

Hundreds of years ago, people believed that some organisms came from nonliving things, such as rotting meat or mud. People had observed, for example, that frogs jumped out of muddy ponds and puddles. Thus they reasoned that the mud produced the frogs. Mice were often found in bags of grain, so people thought that grain produced mice.

An Italian doctor named Francesco Redi questioned this belief. In 1668, he began conducting observations to try to discover what caused maggots to appear on rotting meat. First he put some meat in an open jar. He observed that flies gathered on the meat. When Redi inspected the meat several days later, he observed that there were young flies, called maggots, on the meat.

Redi continued to observe the meat and the maggots for several weeks. After several days, he observed that the maggots stopped moving and became little balls. He placed these balls in an empty jar for further observation. In about a week, Redi observed that the balls broke open and flies emerged from them.

From his observations, Redi wondered whether flies develop from meat or if they appear on meat because other flies lay eggs, which turn into maggots, which turn into adult flies. Based on his observations, Redi made a hypothesis and planned an experiment to test it. A **hypothesis** is a possible answer to a problem or question. Redi's hypothesis was that only flies could produce more flies. His hypothesis could be stated, "If flies can be kept from having contact with rotting meat, then no new flies will appear on the meat."

STEPS IN Making a Hypothesis

Follow these steps to learn how to make a hypothesis.

1 Identify the Question or Problem

Sometimes a problem or a question comes to mind as you observe something. The problem might be an event that you don't understand or a question you have about why something happens.

2 Analyze Your Observations

Think about how to solve the problem or answer the question. Observations you have made or facts you have learned in the past can help you. Use your past observations and experiences to come up with an explanation that could solve the problem or answer the question.

3 State Your Hypothesis

State your explanation in the form of a prediction about what will happen if you perform an experiment to test the prediction. This statement is your hypothesis. An "If . . ., then . . ." format is often used to state a hypothesis. A hypothesis must be able to be tested by an experiment.

1 Redi questioned the common belief that rotting meat produces flies. The possibility that organisms could come from nonliving things was a problem for Redi.

2 Redi made these observations:

- Rotting meat attracted flies.
- Maggots covered rotting meat several days after the flies made contact with the meat.
- The maggots on the rotting meat turned into little balls.
- Flies emerged from the balls after they had been removed from the rotting meat.

Redi analyzed his observations and decided that flies seem to come from rotting meat because other flies lay eggs on the meat, and it takes time for the eggs to develop into flies.

3 Redi's hypothesis was: *If meat is contained so that no flies can lay eggs on it, then no flies will develop on the meat.*

TIP A hypothesis must be able to be tested.

EXAMPLE OF Making a Hypothesis

Read the following passage to see how one student formed a hypothesis.

Observations →

Identify the question or problem →

Observations →

Analyze observations →

Analyze observations →

Hypothesis statement →

Mark was helping make dinner. When he opened the cupboard and reached for the basket of potatoes, he noticed that the potatoes were covered with a green, powdery substance—mold. "Why is there mold all over the potatoes?" Mark wondered.

Mark picked up the basket and carried it to the sink. He noticed that the bottom of the basket was wet. Then he remembered that the last time he had put the groceries away, he had accidentally knocked the basket into the sink. He hadn't bothered to dry off the basket before filling it with potatoes and putting it away. "Could the dampness of the basket have caused the mold to grow?" he wondered. "The potatoes never got moldy when the basket was dry."

"I know how I could find out," Mark thought. "The next time I put away the groceries, if I put some potatoes in a wet basket, I bet mold will grow on them, but not on the potatoes in the dry basket. If potatoes are stored in a damp basket, then mold will grow on them." After the next trip to the grocery store, Mark tested his hypothesis.

USE THIS SKILL

Make a Hypothesis

Read the passage below. Make a hypothesis based on the information in the passage.

Imagine you have an aquarium with many colorful, active fish in it. On the first Saturday of every month, you clean your aquarium. You always follow the same cleaning procedure. First you vacuum the gravel and remove about half of the water from the tank. Then you fill a large bucket with warm tap water, add a few drops of a solution that neutralizes the chemicals in the water, and slowly pour the water into the aquarium.

Last Saturday, you were in a hurry to clean the aquarium so that you could go visit a friend. You vacuumed the gravel, removed the water, and hurried to the sink to fill the bucket with water. But you didn't wait for the water to run warm and instead filled the bucket with cold water. You quickly splashed the neutralizing solution into the bucket and dumped the water into the aquarium.

When you returned a few hours later, you noticed that something about the aquarium was different. The fish weren't swimming around as they usually did. Instead, they were all huddled together at the bottom of the aquarium. "What's wrong?" you wondered. "Why are the fish so still?"

TEST TIP

If you are asked on a test to make a hypothesis, make sure that you consider all the information given. Choose the hypothesis that best matches the information.

Skill 5

HOW TO Collect Data

You Be the Judge

Suppose you and your friends had a contest to see who could jump the farthest. What is the best way to make sure that such a contest is fair?

First you and the other jumpers would observe one another as each person jumps. You would observe if each person began the jump at the start line and also where each person landed. Next the length of each person's jump would be measured. Then the **data,** or information, about the jump would be recorded. In this case, the data is the measurement of the distance of each contestant's jump.

Once everyone had a turn to jump and all the data had been recorded, should the winner be declared? Probably not just yet. Maybe some of the contestants weren't in good positions as they jumped. Someone might have tripped at the last minute, and someone else might have been distracted by a noise. To make sure that the results of the contest are fair and accurate, each person should be allowed to jump several times. When something is tested several times, the results are more likely to be accurate. In science, repetitions of a test to make sure the results are accurate are called **repeated trials.**

STEPS IN Collecting Data

1 Make Observations

Use all your senses to make observations that relate to your investigation. Pay attention to details.

2 Take Measurements

Decide on and use a standard unit of measurement, such as centimeters for length, milliliters for liquid volume, and grams for mass.

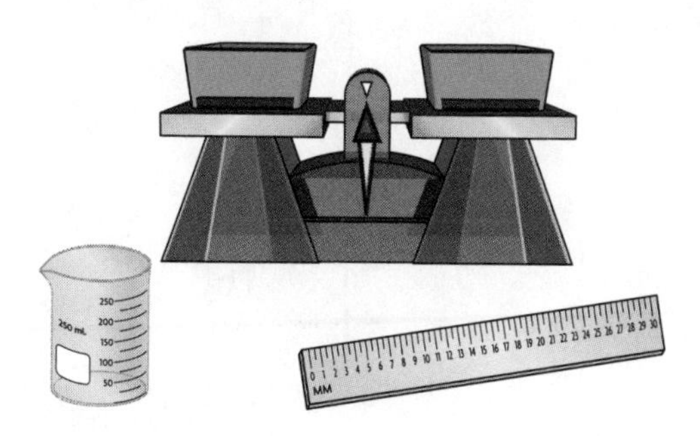

3 Record Data

Data must be accurate, or correct, if you want to to be able draw conclusions from the data. Labeling objects, taking notes, making sketches, and drawing tables are some useful ways of recording data.

4 Repeat the Process

Repeat steps 1–3 at least two more times, recording the data each time. When you have finished all your tests, you can find the average by adding the results together, then dividing this number by the number of times the tests were done.

We each did three trials to see how far we could jump. Emily jumped 143 cm the first time, 162 cm the second time, and 157 cm the third time. To get the average, we added all of Emily's results, then divided that total by the number of trials:

143 + 162 + 157 = 462
462 ÷ 3 = ***154***
(Average distance Emily jumped)

TIP Repeated trials are important when doing science investigations. Anyone else who performs the same experiment should be able to get the same results you did.

EXAMPLE OF **Collecting Data**

The table below shows the data students collected and recorded from a jumping contest. Notice how the students combined the results of repeated trials to come up with a fair and accurate way to determine the winner.

Results of the Broad-Jump Contest

Distance Jumped (in centimeters)

Name:	1st Jump +	2nd Jump +	3rd Jump =	Average ÷ 3 =
Jamal	181	187	190	186
Sean	128	142	147	139
Anna	166	160	172	166
Kim	158	164	149	157
Jacob	186	180	171	179
Heather	145	152	159	152
Pablo	190	186	194	190
Sara	178	182	189	183

Data: Measurement of each jump

Repeated trials

Average of all trials

USE THIS SKILL

Collect Data

Read the paragraphs and answer the questions below.

Ben's soccer coach wanted all players to test themselves to find the average distance they could kick the soccer ball. Ben took his ball to the practice field behind the school, where the whole length of the field was marked in meters. Ben set the ball at the "zero" line and kicked as hard as he could. The ball went down the field to the 13-meter mark. Ben wrote the date and the distance in his notebook.

The next day, Ben went back to the practice field. He set the ball at the "zero" line and kicked as hard as he could. This time, the ball went to the 18-meter mark. Ben decided to test his kick two more times, then average the three distances. The second kick, the ball went only 14 meters. The third kick, the ball went all the way to the 22-meter mark. Ben recorded the data from each kick, then added the three numbers. He divided the total by three, the number of times he had tested his kick distance that day, to get his average kick distance.

1. Make a chart and record all the data about Ben's kicks.
2. How far did Ben kick the ball on the first day?
3. What is the farthest Ben kicked the ball on the second day?
4. What was the average distance Ben kicked the ball on the second day?

TEST TIP

If you are asked to collect data for a test, make sure you use the correct measurements and record numbers accurately.

Skill 6

HOW TO

Control Variables

The Period of a Pendulum

Ever since Galileo first observed the regular motions of a swinging lamp in the 1580s, people have been fascinated by the movement of pendulums. A **pendulum** is an object that is hung from a fixed point with a string or chain and allowed to swing freely. The hanging object is called the **bob.**

Galileo and other scientists studied the motions of pendulums and found that they are related to many other types of motion. Pendulums have been used to measure increases in speed due to gravity at different places, to explain the rotation of Earth on its axis, and to develop the first reliable clocks.

To set a pendulum in motion, you pull back the bob and then let it go. The bob will move past the bottom position and back up. Then it will briefly stop, reverse its direction, and swing back toward the position where you first let it go. The pendulum will continue to swing back and forth. The time it takes a pendulum to swing over and back once is called the **period** of the pendulum.

Pendulum

Period of a pendulum

As you might guess, different pendulums have different periods. What can affect the period of a pendulum? To find out, you could make a pendulum and then conduct an experiment to see how different variables affect the period of the pendulum. **Variables** are the factors that could change the results of an experiment. Variables affecting the period of a pendulum include the length of the string or chain and the distance from which a pendulum's bob is dropped and set into motion.

During an experiment, it is important to change only one variable at a time. All the other variables are kept the same, or controlled. When you change only one variable at a time, you are making sure that any change during the experiment is the result of that one variable you changed. In the case of the pendulum, if you changed both the height from which you dropped the bob and the length of the string, you would find that these changes affected the pendulum's period. However, you would not know how each variable caused the period to change.

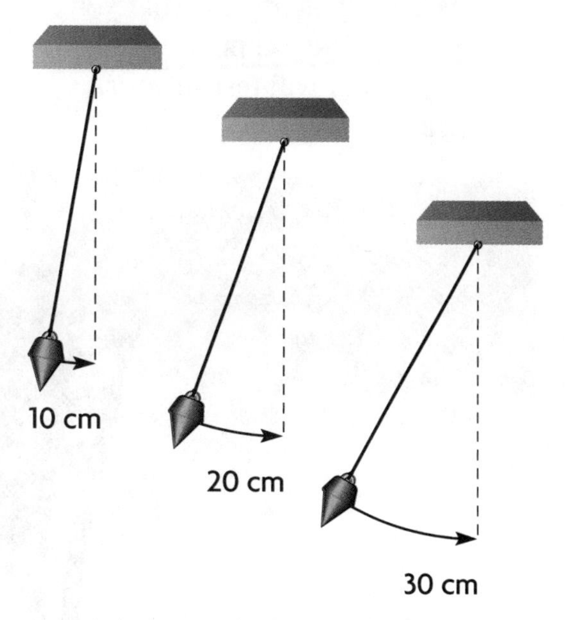

Pendulums with three different distances of movement

STEPS IN Controlling Variables

Follow the steps below to learn how to control variables.

1 Identify the Question

First clearly define and state what you will try to find out in your investigation.

I will find out if the mass of the bob has an effect on the period of a pendulum.

2 Identify What You Will Test

Once you have identified what you want to find out, you must focus on what factor you will test in your investigation.

The variable that I will test is the mass of the bob. I will use different masses to see if they change the pendulum's period.

3 Identify Other Variables

Think about all the other factors, or variables, that might affect the results of your investigation.

Other possible variables:

- *Length of the pendulum's string*
- *Distance that the bob is pulled back*

TIP **If you cannot immediately identify the factor that you will test, you should rethink or rephrase your question.**

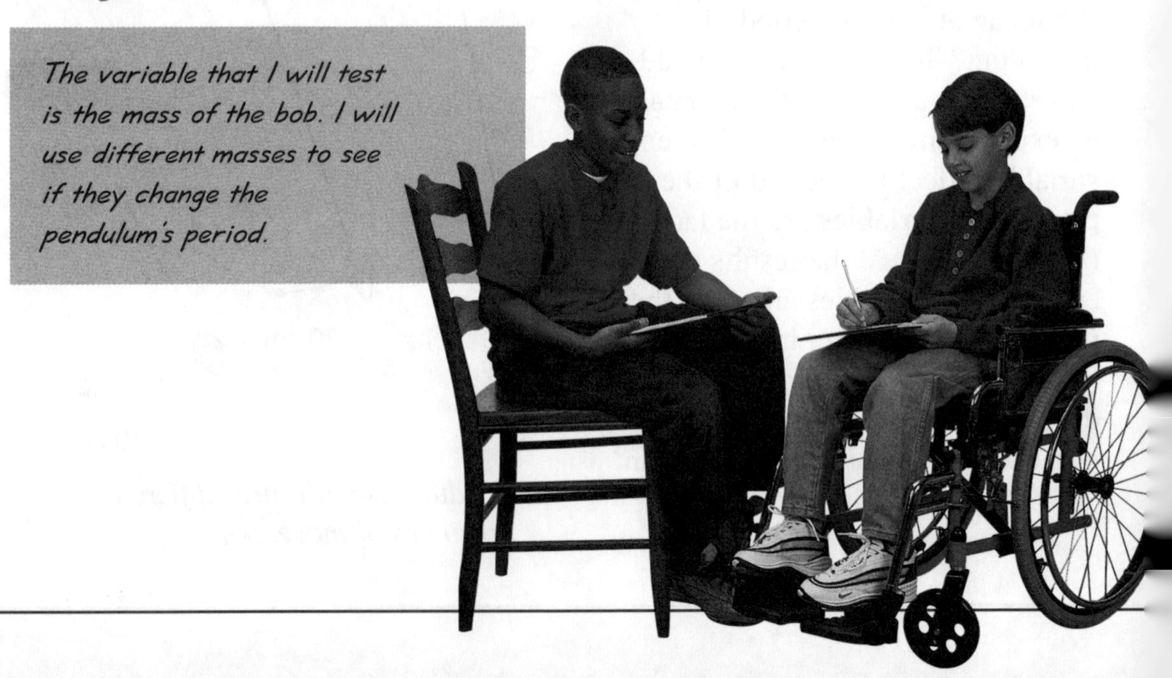

4 Control the Variables

Once you have identified all the variables, you must make sure that the only variable you change is the one that you are testing. All the other variables must stay exactly the same.

Throughout the investigation, use exactly the same length of string.

Throughout the investigation, make sure the bob is released from exactly the same distance as it was before the mass of the bob was changed.

5 Repeat Your Test

Perform several trials to make sure that your results are accurate. For each trial, change only the one variable that you are testing. Measure and record the results of each one of your trials.

Data and Observations

Trial	Mass of Bob (Number of Metal Washers)	Pendulum Period (Seconds)
1	1	2
2	2	2
3	3	2

EXAMPLE OF Controlling Variables

Read below to see how one student did an investigation by changing only one variable at a time.

Identify the question → Jordan decided to do an investigation to find out what variables affect the period of a pendulum. For his **Identify what to test** → investigation, Jordan decided to test one variable: the length of the string holding the bob. His question was: "If the length of the string holding the bob is changed, will the period of the pendulum also change?" He **Identify other variables** → thought that the other variables of the experiment could be the mass of the bob and the distance that the bob was pulled back before being let go. Jordan planned to control these variables by using the same bob and the same distance from which it is raised and let go for all of his trials.

During his experiment, Jordan made four trials. For each trial, he changed only the length of the string. He **Control variables** → made sure the other variables were exactly the same for each trial. On the first trial, **Repeat tests** → he hung the bob from a 10-cm piece of string; on the second trial he hung the bob from a 15-cm piece of string; on the third trial he hung the bob from a 20-cm string, and on the fourth, a 25-cm string. At each trial, Jordan measured and recorded the period of the pendulum.

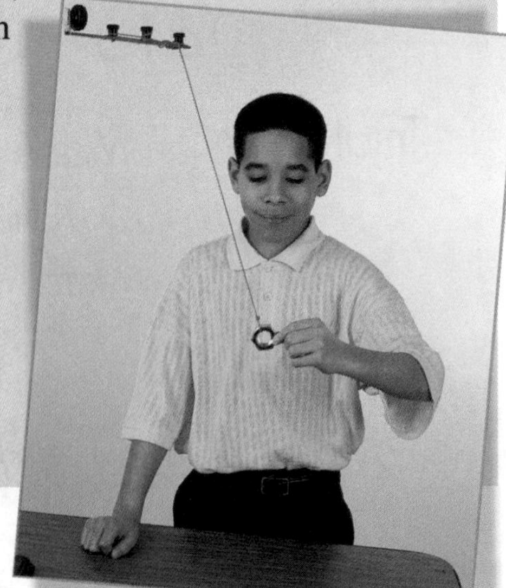

USE THIS SKILL

Control Variables

Examine the illustration below, which shows the three trials of an experiment related to the motion of a pendulum. Identify the variable being tested. Also identify the variables that are being controlled.

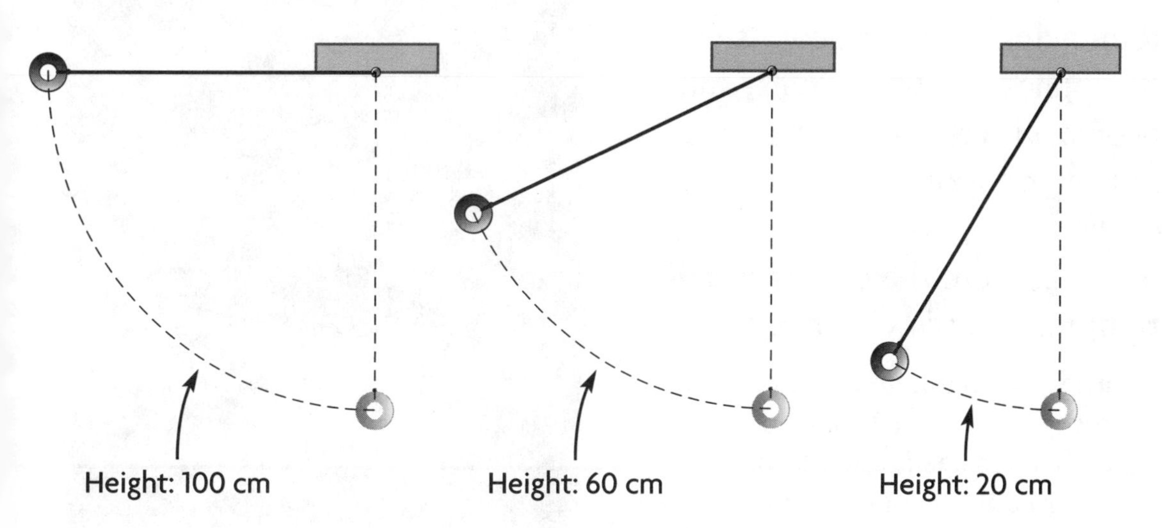

The first pendulum was raised and released from a height of 100 cm and had a period of 3 seconds.

The second pendulum was raised and released from a height of 60 cm and had a period of 2 seconds.

The third pendulum was raised and released from a height of 20 cm and had a period of 1 second. All three have a string that is 100 cm long and the same bob size.

TEST TIP

A science test may ask you to identify variables in an example experiment. You might find it helpful to write a list of possible variables and then cross them off your list as you identify each one.

Skill 7

HOW TO

Design an Experiment

Acid Rain

Rain helps nature renew itself. It supplies plants, animals, and people with the water we need to live and grow. In many areas of the world, however, the water that falls from the clouds is **acid rain,** rain that is very acidic.

An **acid** is a compound that releases hydrogen ions when dissolved in water. Some acids are weak, but others are quite strong. Tomatoes, lemons, and vinegar contain weak acids, whereas car batteries contain a much stronger acid. The strength of an acid can be measured by the pH scale. The *pH* stands for "potential for hydrogen." The scale goes from 1 to 14. The number 7 on the pH scale is the middle point: it indicates that a solution is not acidic. Pure water has a pH of 7—it is described as neutral. As you go lower down the pH scale, a solution becomes more acidic. Normal rainwater is only slightly acidic, with a pH of about 5.6.

As water from Earth's surface evaporates, it rises into the air. Carbon dioxide naturally mixes with water vapor in the air as it forms raindrops. Pollutants are released into the air by the burning of fuels such as gasoline and coal. These pollutants react with water vapor in the air to form strong acids. This makes rain more acidic than normal.

Acid rain is harmful to many life forms. Acid rain can damage pond and lake ecosystems by killing water plants and animals. Other animals that depend on these organisms for food then begin to die also. Acid rain also washes away nutrients from soil. This can weaken and kill trees and other plants. Because of these harmful effects, many people are trying to limit pollution that causes acid rain, as well as to clean up areas that have been harmed by acid rain.

The effects of acid rain were identified by scientists who made observations and formed hypotheses. These scientists then designed and conducted experiments to test the hypotheses they had made. If you wanted to learn more about the effects of acid rain, you could make observations, form a hypothesis, and then design and conduct an experiment.

Acid rain is harmful to living things.

STEPS IN Designing an Experiment

Follow these steps to learn how to design an experiment.

1 Develop the Procedure

Once you have made a hypothesis, you should know what you want to test. The procedure tells exactly what you will do to test your hypothesis. Write your procedure as a list of steps that you will follow when doing your experiment. Identify any variables that might affect the results of your experiment. Explain how you will control these variables.

Materials:

30 g of limestone chips
200 mL of vinegar
2 250-mL beakers
stirring rod
balance
one piece of filter paper
paper towels
water

Procedure:

1. Observe the limestone chips. Note whether their surfaces are smooth or rough.
2. Put the limestone chips in a 250-mL beaker.
3. Add 200 mL of vinegar to the limestone chips.
4. Stir the chips in the vinegar for a few seconds, then let sit five minutes.
5. Carefully pour most of the vinegar out of the beaker. Rinse the chips in water and allow them to dry on paper towels.
6. Observe the surfaces of the limestone chips.
7. Measure the mass of the chips and compare their mass now with their mass before.

2 Develop a Materials List

Think about the materials you will need to conduct your experiment. In addition to the items being tested, list any containers and tools you will need. Be specific about the sizes and amounts of materials.

3 Review the Procedure

A good experiment must be able to be repeated by you or anyone else who reads your materials list and procedure. Before you begin your experiment, it is important to think through your procedure. You may find that you need additional materials or different amounts of materials. Reviewing both your hypothesis and your procedure will help you conduct your experiment successfully.

4 Conduct Your Experiment

Once you have followed these steps, you are ready to begin doing your experiment. Make sure to follow all class safety rules and ask an adult for help.

TIP You will need to do your experiment several times to get accurate results.

EXAMPLE OF Designing an Experiment

Read the following passage to see how one student designed an experiment.

Maria designed an experiment to see what level of acidity tomato plants could tolerate before they were damaged. She hypothesized that small increases in acidity would not hurt the tomato plants.

Develop procedure

Here is what I will do:

1. *I'll plant six tomato plants that are the same age in separate containers. To control the variables of soil type and amount of sunlight, I'll plant them in the same kind of soil and keep them all in the same window.*
2. *To control the variable of amount of water, I'll give each plant exactly the same amount of water.*
3. *Every morning, I'll spray different strengths of vinegar solution on each of the six plants.*
4. *I'll observe the plants every day to see how healthy they are. I'll record these observations, plus the exact measurements of the vinegar solutions I used.*

Within three or four weeks, I should find out just how much acid is harmful to the plants.

Develop materials list

Maria wrote a list of what she would need to conduct this experiment.

Maria thought through her experiment design and wrote her hypothesis. She reviewed these several times, made a few changes, and then began her experiment.

Review procedure

Conduct experiment

USE THIS SKILL

Design an Experiment

How acidic is the rain in your area? Read the information and study the materials list below. Then design an experiment that will answer that question.

An indicator is a substance that changes color according to the level of an acid in a solution. You can use an indicator, along with a color chart, to find the pH of a solution. Remember that a pH of 7 is neutral, and that as the acidity of a solution increases, the pH number decreases.

Materials List
- large, clean bucket
- pH indicator solution
- pH color chart
- 6 250-mL beakers
- sticky labels
- pens and markers
- notebook or notepaper

TEST TIP

Some science tests may give you a hypothesis and ask you to write a plan for an experiment to test it. Be sure the experiment you design is a good test of the hypothesis. Your materials list and procedure should relate only to the suggested hypothesis. Remember that a simple test is usually better than one that is complicated.

Skill 8

HOW TO Draw Conclusions

How Do Plants Grow?

When you think of a plant growing in soil, you probably picture the roots growing down and the stems and leaves growing up toward the sky. Have you ever wondered how plants grow this way?

Plants are able to respond to some conditions within their environment. Plants can respond to stimuli such as light, water, and the force of gravity. A plant's response to a stimulus is called a **tropism.**

Oscar wanted to find out if the position that seeds are planted in soil affects the direction that the roots grow. He did an experiment to find out. He first made a hypothesis, then set up and carried out his experiment. From the results of his experiment, he was able to draw a conclusion. A **conclusion** is an explanation of the results of an experiment. A conclusion will tell whether or not the data collected during an experiment support the hypothesis.

Oscar's hypothesis: If bean seeds are planted in different directions, then the roots will grow in the directions in which the seeds are planted.

Oscar's experiment: 20 bean seeds were planted along the side of four separate plastic cups. Five of the beans were planted pointing up, five were pointing down, five were turned to the left, and five were turned to the right. All had equal amounts of light and water.

STEPS IN **Drawing a Conclusion**

Follow these steps to learn how to draw a conclusion from the results of an experiment.

1 Think about What Happened

Think about what you observed during your experiment. Carefully consider what your data show.

2 Write a Conclusion Statement

Write a statement to explain what happened. This explanation will be your conclusion. It summarizes what you have learned from the experiment.

3 Review Your Hypothesis

Compare your conclusion to your hypothesis. The conclusion might support the hypothesis. This means that your idea about what would happen in the experiment was correct. However, the conclusion might not support the hypothesis. This means that what you thought would happen did not. Don't worry if the conclusion does not support your hypothesis. Both kinds of results are important when doing science experiments. Doing an experiment may raise new questions or issues that need to be investigated. Some results can give you clues about how to change your hypothesis and start over.

- All the seeds except one grew roots.
- All the roots grew down.
- **Conclusion:** The roots grew down no matter which direction the seeds were planted.
- The conclusion did not support my hypothesis.
- A new question I have is: Why didn't that one seed grow roots?

TIP **Sometimes you may need to repeat an experiment more than once before you can draw a conclusion.**

EXAMPLE OF Drawing a Conclusion

Read the following passage to see how one student drew a conclusion based on the results of an experiment.

Hypothesis → *What I think:* A plant will grow toward light even if obstacles block the path to the light.

Experiment → *What I did:* I put a young plant in one end of a shoebox with two partitions and one hole to allow light in to only the far end of the box. The box was kept closed. The lid was removed for a few seconds once every three days to water the plant and record observations.

Observations → *What I observed:*

Week 1 Week 2 Week 3

Conclusion statement → *What happened:* The plant grew around the cardboard partitions toward the hole where sunlight came into the box.

Review the hypothesis → *This experiment supported my hypothesis.*

USE THIS SKILL

Draw a Conclusion

Read the paragraph and examine the photographs below. Draw a conclusion statement about the results and decide if the conclusion supports the hypothesis.

Carli noticed that her shamrock plant was bending toward the light coming in through a window. She decided to move the plant to see what would happen. The pictures below show what happened.

Carli's hypothesis: If I turn the plant away from the light, the stems will grow back toward the light.

Monday, 9:00 A.M.

Monday, 10:00 A.M.

Tuesday, 9:00 A.M.

TEST TIP If you are asked on a test to draw a conclusion, make sure that you base your conclusion on the facts provided.

Skill 9

HOW TO

Write an Observation Report

Crystal Clear

Have you ever heard that each snowflake is different from every other snowflake? This is true, but all snowflakes are still recognizable as snowflakes. Why? Because snowflakes almost always have the same pattern: a hexagon.

Each snowflake is a **crystal,** a solid in which the atoms form repeating geometric patterns. Ordinary table salt is also a crystal. If you looked at salt with a magnifying lens, you would observe that each grain of salt is cubic. The groups of atoms within each grain of salt are arranged so that they form a cube.

The formation of crystals, called **crystallization,** can happen in several different ways. Many mineral crystals form when hot, melted rocks cool and solidify. Minerals often form beautiful crystals of different colors and shapes.

Mineral crystals

When minerals are dissolved in a liquid and the liquid evaporates, the atoms in the molecules of minerals are left behind and form crystals. Huge deposits of salt and other minerals, called salt flats, form in this way.

Salt flat

Crystals can be formed in a process called seed crystallization. In seed crystallization, a crystal of a substance is hung in a solution made of the same substance and water. Other crystals then begin to form on this "seed" crystal. Eventually the crystals will form a large crystal that has the same shape as the seed crystal.

If you decided to investigate the process of seed crystallization, you could actually observe the growth of a crystal over a period of several weeks. To summarize your observations and draw conclusions from them, you could record them in an observation report. An **observation report** is a written record of what you have seen during a science investigation. A science observation report is written at the end of an investigation. The report gives information about all the steps involved in an investigation: the problem, the hypothesis, the materials, the procedure, your observations of the results, and your conclusion.

For a science observation report, you do not always have to write in full paragraphs, as you do for many other kinds of reports. Each section is introduced with a bold heading. To write an observation report, record your information in the format shown on the next two pages.

STEPS IN Writing an Observation Report

1 Determine the Problem

Write the problem that your investigation was designed to solve or the question that it was designed to answer. Your problem should be in the form of a question.

2 State Your Hypothesis

Write your hypothesis statement. Remember that a hypothesis is a prediction of what you expected to happen as a result of your investigation.

3 Describe Your Procedure

The procedure section is a description of the methods that you used in your investigation. Begin the procedure section with a sentence that tells what you wanted to find out in your investigation. Then describe the materials that you used and the steps that you followed. Be specific about the sizes and amounts of materials that you used and about the length of time needed for each step.

4 Record Observations

The observation section is a detailed description of what you saw over the course of your investigation. Your observations should be recorded in the order in which you observed them. Describe all the details of what you observed, including colors, shapes, sizes, and movements.

5 Draw Conclusions

The conclusion section sums up your observations and tells what you have learned from your investigation.

TIP Although most of us rely mainly on our sense of sight during an investigation, our other senses also help us make observations. Include any sounds, smells, and changes in textures or temperatures that you observe during an investigation if they are relevant.

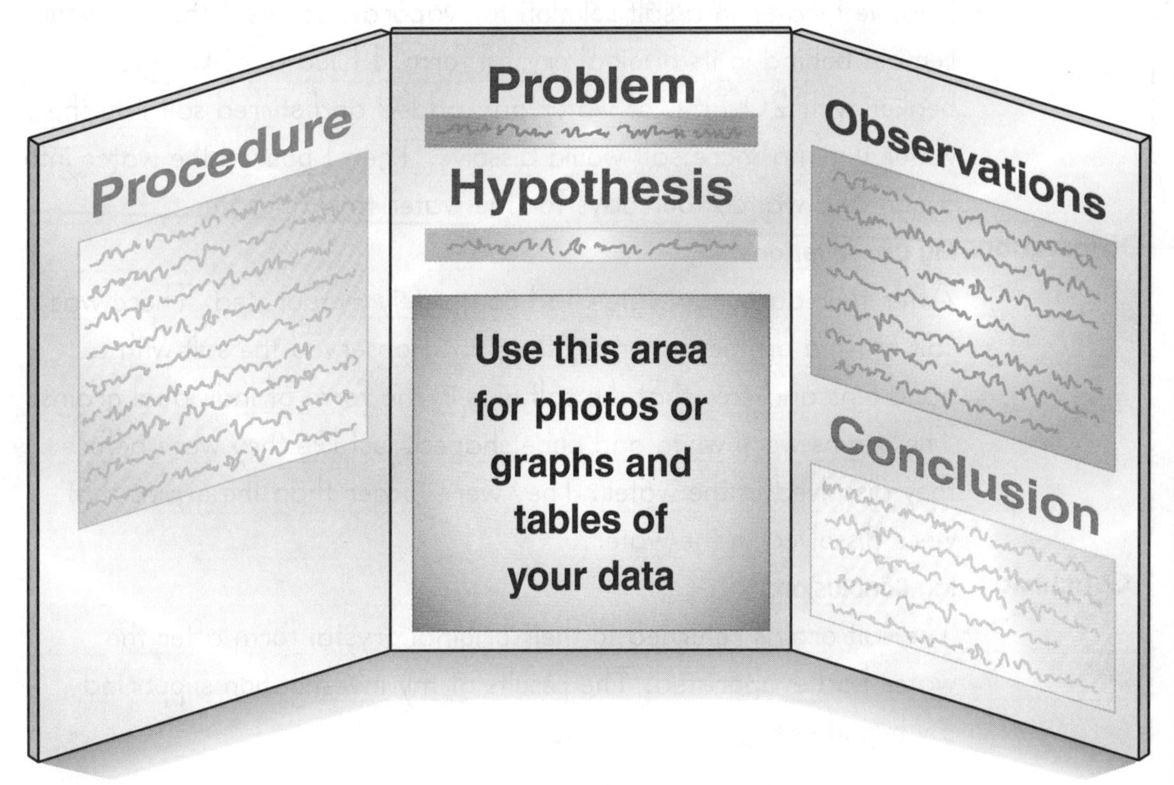

An observation report can be used for a science fair project.

EXAMPLE OF **Writing an Observation Report**

Read the observation report below that one student wrote after completing a simple investigation.

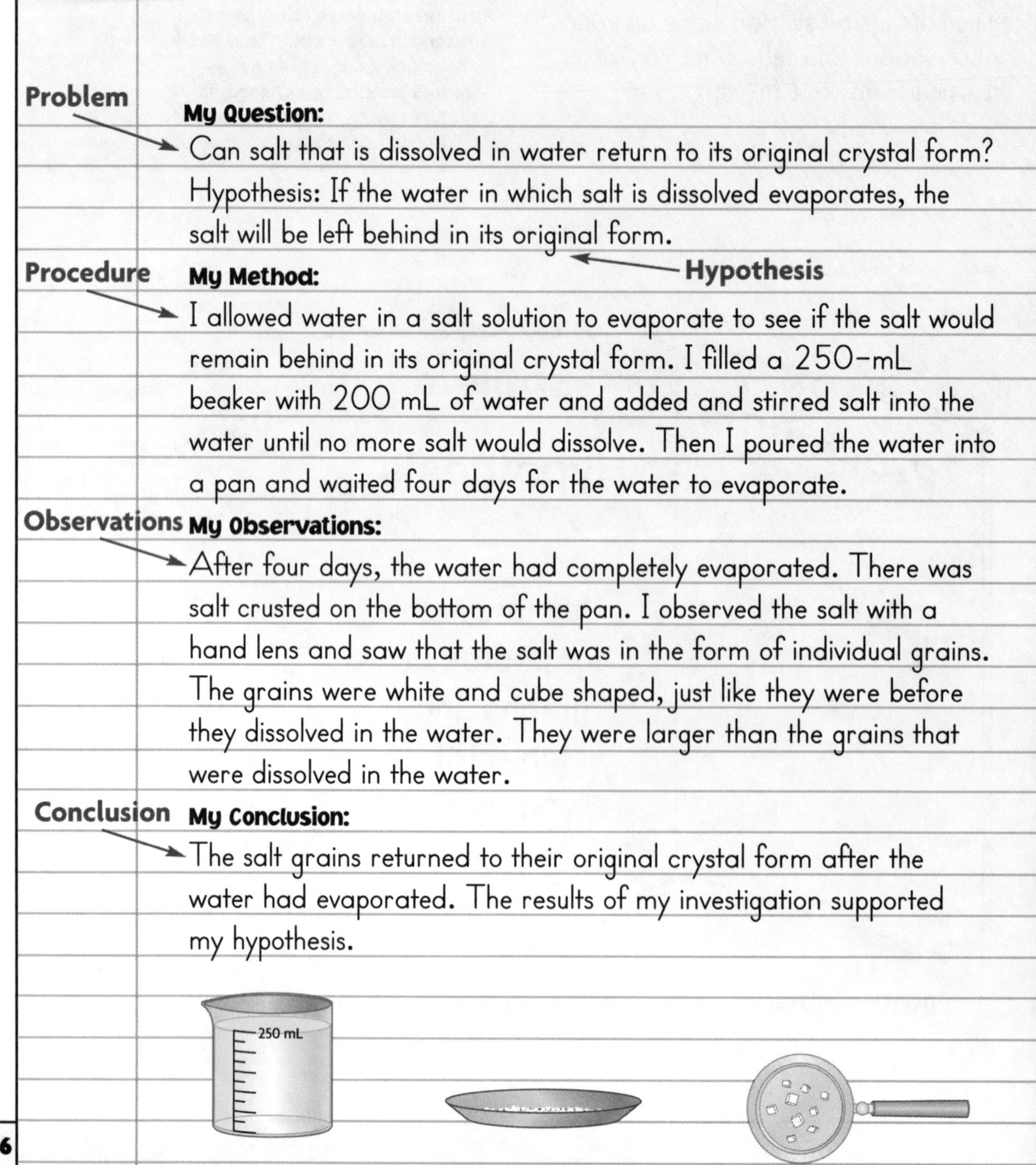

Problem

My Question:

Can salt that is dissolved in water return to its original crystal form?

Hypothesis: If the water in which salt is dissolved evaporates, the salt will be left behind in its original form.

Hypothesis

Procedure

My Method:

I allowed water in a salt solution to evaporate to see if the salt would remain behind in its original crystal form. I filled a 250-mL beaker with 200 mL of water and added and stirred salt into the water until no more salt would dissolve. Then I poured the water into a pan and waited four days for the water to evaporate.

Observations

My Observations:

After four days, the water had completely evaporated. There was salt crusted on the bottom of the pan. I observed the salt with a hand lens and saw that the salt was in the form of individual grains. The grains were white and cube shaped, just like they were before they dissolved in the water. They were larger than the grains that were dissolved in the water.

Conclusion

My Conclusion:

The salt grains returned to their original crystal form after the water had evaporated. The results of my investigation supported my hypothesis.

USE THIS SKILL

Write an Observation Report

Read the paragraph and follow the instructions below to practice the steps in writing an observation report.

John conducted an investigation to learn about the process of seed crystallization. Study John's problem statement, his hypothesis, and the materials list for his investigation. Then write procedure, observation, and conclusion sections that could be used to complete John's observation report.

Problem: *How can crystals form from a solution?*

Hypothesis: *If a copper sulfate crystal left from the evaporation of water is hung by a thread in the original solution, then a larger crystal will form.*

Materials

- *1 L of water*
- *1 kg of copper sulfate*
- *1 large glass jar with lid*
- *stirring rod*
- *shallow dish*
- *tongs*
- *50-cm length of thread*
- *tape*

TEST TIP

You may be asked to record your observations on a science test. Make sure the measurements and other data you collect are accurate and in the correct format.

Reading and Thinking Skills

Skill 10 HOW TO Classify

Skill 11 HOW TO Compare and Contrast

Skill 12 HOW TO Determine Cause and Effect

Skill 13 HOW TO Tell Fact from Opinion

Skill 14 HOW TO Find the Main Idea

Skill 15 HOW TO Take Notes

Skill 16 HOW TO Predict

Skill 17 HOW TO Make Generalizations

Skill 18 HOW TO Make a Decision

Skill 19 HOW TO Work in a Group

Skill 10

HOW TO

Classify

The Stars in the Sky

On a clear night when you look up in the sky, you can see the light from stars. Some of these stars are in the same galaxy as Earth—the Milky Way galaxy. Others are in different galaxies across the universe. One star in the Milky Way galaxy is the sun. But what does the sun have in common with the other 200 billion stars in this galaxy?

Stars are huge balls of hot gas. They give off energy in many ways, including heat, light, and radio waves. All stars are huge, but they can vary in size. When scientists study the sun and other stars, they look at different features such as the size of the star. They might also look at characteristics such as temperature, brightness, or distance from Earth.

When scientists **classify** stars, they put them into groups based on certain characteristics. For example, we know that stars burn and create energy. If we look at the temperatures of stars, we can see that some stars are hotter than others. This feature can be divided into two groups: hot stars and cool stars. Another feature—the color of stars—can be used to break these two groups into subgroups. The color of a star shows how hot it is.

When you classify items, you put them into groups based on what they have in common. Classifying helps organize information so it can be easily understood. You can learn how to classify by following the steps on the next page.

STEPS IN Classifying

1 Analyze Your Items

Look over the items you want to classify. What common features do these items have? Think about what kinds of classifying categories you could use to sort the items, such as size, shape, or color. To create your own way to sort items, keep in mind your overall purpose as you consider what similar features are important.

Stars

- Huge clouds of burning gas held together by gravity
- Create heat and light through nuclear fusion

Features

Hot Stars

Temperature is greater than 10,000 °C

Cool Stars

Temperature is less than 10,000 °C

Groups

TIP When thinking about how to classify your items, **compare and contrast** your items. How are they alike? How are they different?

2 Select One Shared Feature

Choose a feature that many of your items share. Place items with this feature into a group. Put the other items without this feature into a second group.

3 Repeat the Process

Check each group to see if there is another feature that might make a subgroup. If so, divide the items into smaller groups. Continue until every item is classified according to as many useful categories as you can find.

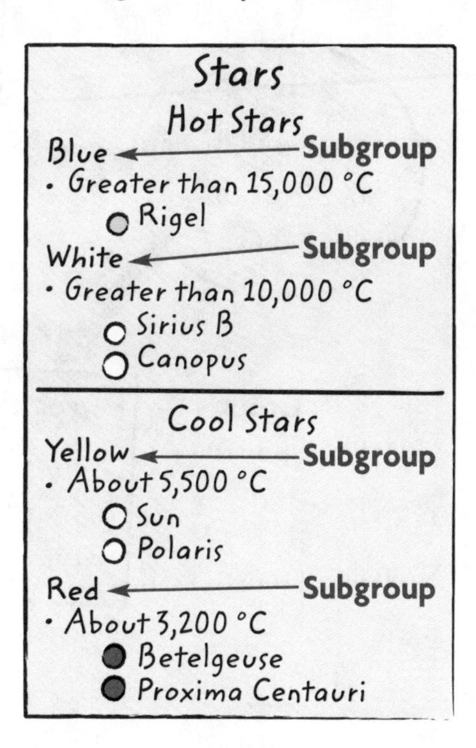

Stars
Hot Stars
Blue
Subgroup
• Greater than 15,000 °C
Rigel
White
Subgroup
• Greater than 10,000 °C
Sirius B
Canopus
Cool Stars
Yellow
Subgroup
• About 5,500 °C
Sun
Polaris
Red
Subgroup
• About 3,200 °C
Betelgeuse
Proxima Centauri

EXAMPLE OF Classifying

Look below to see how one student began to classify different groups of stars, called galaxies. Galaxies are huge families of stars, gas, and dust held together by their own gravity. Think about the characteristics of each group.

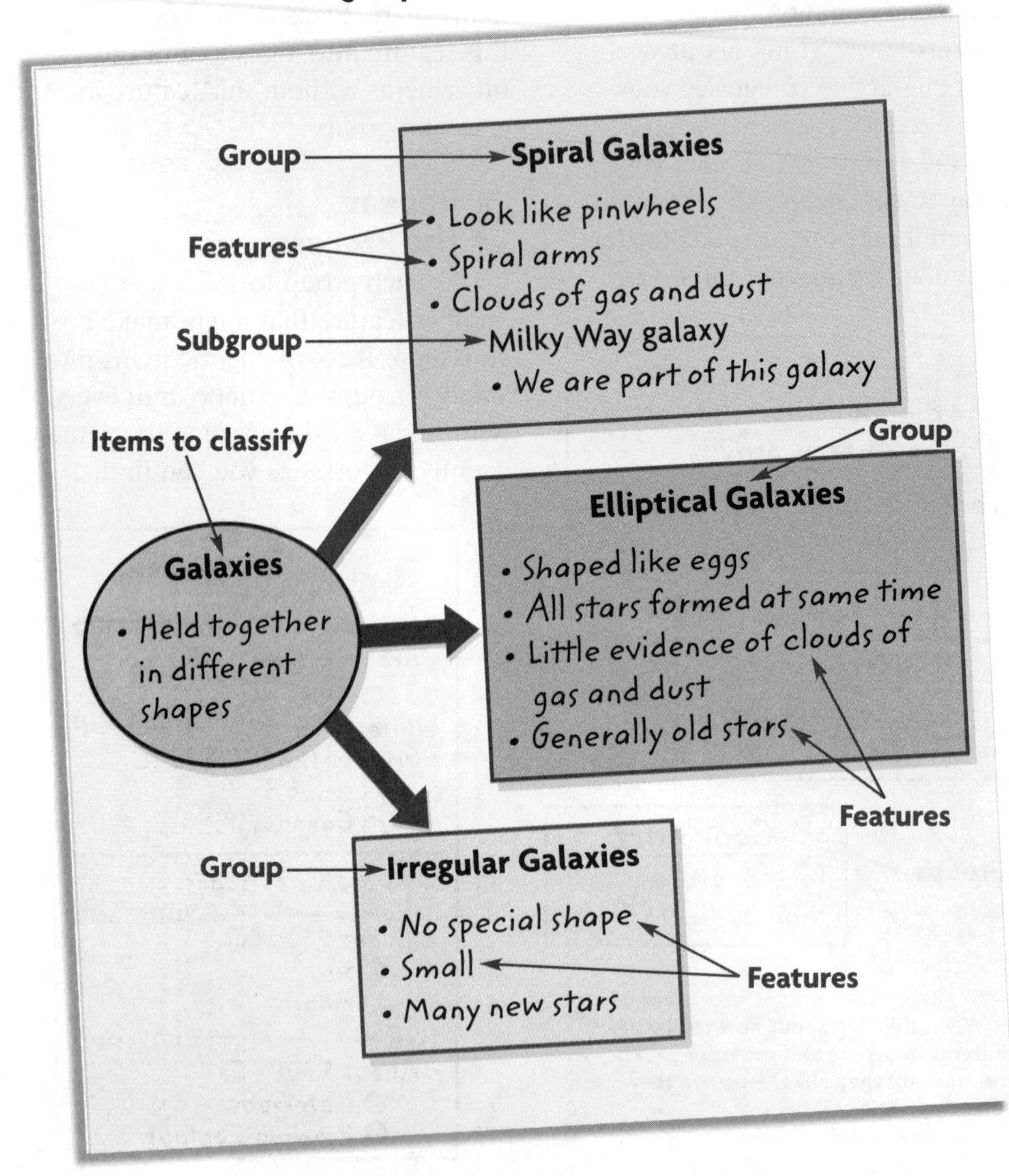

USE THIS SKILL

Classify

Look at the star characteristics provided in the chart below. On another sheet of paper, classify these stars in as many different ways as you can. You may want to use charts, lists, or other ways to organize the information as you classify.

Star Name	Star Type	Distance from Earth (in light years)
Sun	yellow	less than 1
Proxima Centauri	red dwarf	4.2
Alpha Centauri A	yellow	4.3
Lalande 21185	red dwarf	8.1
Canopus	white supergiant	200
Rigel	blue/white supergiant	910
Achernar	blue/white	85
Arcturus	red giant	36
Sirius B	white dwarf	8.6

TEST TIP **You may be asked to classify words on a test. Look them over to see what most of them have in common. Ask yourself which one does not belong.**

Skill 11

HOW TO

Compare and Contrast

Muscles in Motion

You are probably aware of the muscles in your arms and legs. Did you know that you also have muscles in your heart, intestines, and blood vessels? You use these muscles every day, but you cannot see them.

Your body has about 600 muscles. The ones that you can control are called **voluntary muscles.** Smiling, talking, and walking to school are examples of actions you control by using voluntary muscles. Muscles that are not under your control are called **involuntary muscles.** These include the muscles that help you breathe, digest food, and pump blood throughout your body. Involuntary muscles allow the organs of your body to function without your ever thinking about them. Humans and animals need both kinds of muscles to survive.

Muscles can also be grouped by other characteristics. A muscle can fall into one of three categories. It can be a skeletal, smooth, or cardiac muscle. Skeletal muscles are attached to your bones and allow them to move. Smooth muscles control movements inside your body, such as the digestive process. Cardiac muscle makes the pumping action of the heart possible.

Comparing and contrasting can help scientists learn more about the body's muscles. When they **compare,** they look for qualities that are alike among two or more things. When they **contrast,** they look for qualities that are different.

STEPS IN Comparing and Contrasting

Like a scientist, you can use comparing and contrasting to help you understand new concepts.

1 Focus on Your Topics

Think about what you will be comparing and contrasting. Learn as much as you can about your topics. Write the basic facts about each topic.

2 Organize Information

Organize the facts in a way that will help you see the ways the topics are the same and different. Graphic organizers such as Venn diagrams and charts are very helpful when organizing information.

- Use a chart when comparing and contrasting three or more things. Across the top of the chart in different columns, list each of the things you wish to evaluate. Down the left side of the chart, list the characteristics by which you are evaluating each item. Then fill in the chart with information about your subjects.
- To make a Venn diagram, draw two large, overlapping circles. Label each circle. Write the details that are different in each circle. List the details that are the same for both in the overlapping area. This helps you see similarities and differences between the two subjects at a glance.

Venn diagram

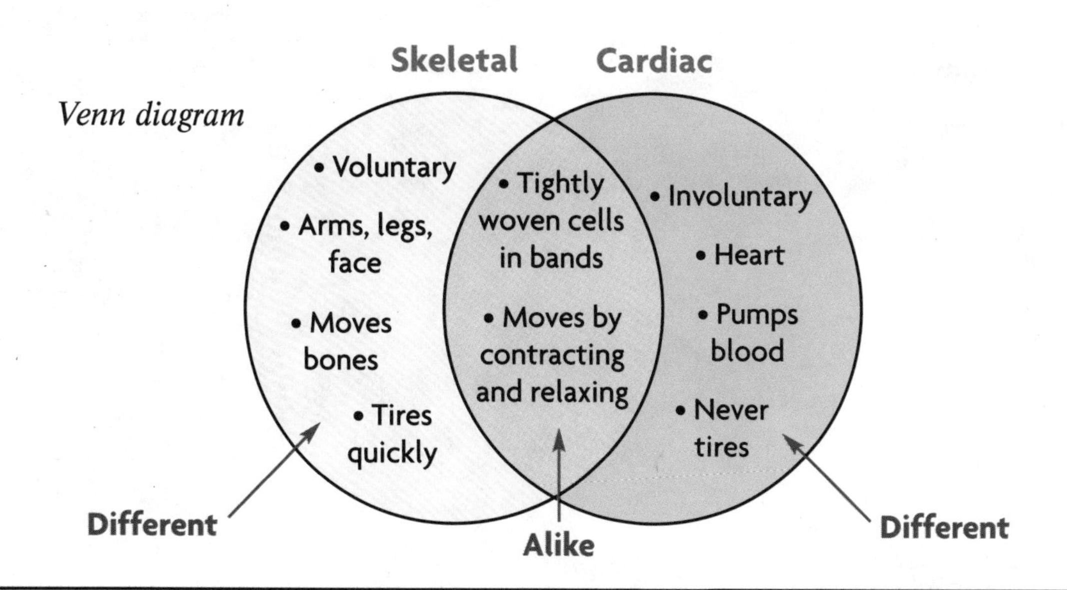

EXAMPLE OF Comparing and Contrasting

Read the chart below to see how a graphic organizer can be used to help compare and contrast three types of muscles.

Characteristics of muscles

Types of muscles

Types of Muscles

CHARACTERISTIC	SKELETAL	SMOOTH	CARDIAC
Voluntary/ Involuntary	Voluntary	Involuntary	Involuntary
Description	Tightly woven into bands	Long, slender shape	Tightly woven into bands
Location	Arms, legs, face	Walls of blood vessels, stomach, esophagus, intestines	Heart
Function	Moves bones	Controls involuntary movement in body	Pumps blood
Movement	Contracts and relaxes	Contracts and relaxes forming a wavelike action	Contracts and relaxes
Reaction Time	Quick	Slow	Repeated contractions
Fatigue Time	Quick	Never tires	Never tires

USE THIS SKILL

Compare and Contrast

Read the following passage about skeletal muscles and smooth muscles. Compare and contrast these two types of muscles using a graphic organizer to show how they are alike and different.

Muscles of the Body

The muscles of your arm allow you to throw a ball, lift a book, or brush your teeth. Because you can control them, they are called voluntary muscles. Arm muscles are also called skeletal muscles because they move bones. They do this by working in pairs that are opposite of each other. For example, as the biceps on the top of the upper arm contract, the triceps at the back of the upper arm relax. The cells of the arm muscles are tightly woven together to form bands. The muscles of the arms and legs react when given a signal. However, the arm and leg muscles tire out and need a rest when doing hard work.

The muscles of the digestive system are smooth muscles. Because they function without your having to think about controlling them, they are considered involuntary muscles. The cells of your digestive system's muscles are long and slender. These muscles contract and relax to form a wavelike action that moves digesting food along inside the intestines. Smooth muscles respond slowly, not quickly like skeletal muscles. Smooth muscles do not tire out as arm and leg muscles do.

TEST TIP

You may be asked to answer compare and contrast questions on a test. Read the test questions very carefully, making sure you identify exactly what you must compare or contrast before you begin to think about your answer. Then decide how these things are alike or different.

Skill 12

HOW TO

Determine Cause and Effect

Getting Fit

Have you ever noticed how much better you feel after a long walk or bike ride? You feel more alert, relaxed, and happy. What is the cause of these positive effects?

Scientists have learned that regular aerobic exercise causes a variety of health benefits. **Aerobic exercise** is rhythmic exercise that increases the heart rate and strengthens the heart and lungs. Running, cycling, and swimming are all aerobic exercises.

To understand more about the human body, scientists try to determine the causes and effects of different actions and events. A **cause** is the reason something happens. That which happens as a result is the **effect.** For example, your heart rate increases when you play soccer. The exercise of playing soccer is the cause, and the increased heart rate is the effect.

Sometimes one cause brings about several effects. Identifying causes and effects will help you understand why events occur. As you read the following paragraph, look for one cause that results in multiple effects.

Marta has been jogging for 30 minutes three times a week. As a result, her heart and lungs have become more efficient at pumping oxygen throughout her body. Marta's leg muscles are stronger, and she has less body fat. Marta has also been feeling happier and less stressed. For all these reasons, Marta will continue her jogging program.

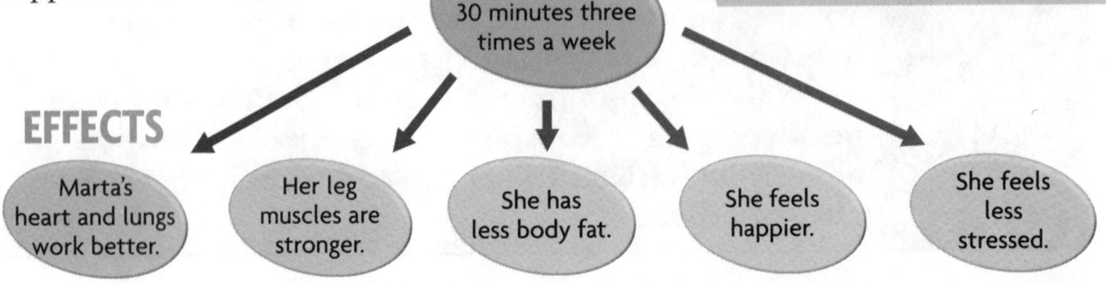

CAUSE

Jogging for 30 minutes three times a week

EFFECTS

- Marta's heart and lungs work better.
- Her leg muscles are stronger.
- She has less body fat.
- She feels happier.
- She feels less stressed.

STEPS IN Determining Cause and Effect

Follow the steps below to identify a single cause that has many effects.

1 Look for Clue Words

When one event causes several effects, you will notice clue words. Clue words can help you find cause-and-effect relationships in your reading. Some of these words are *as a result, because, caused, consequently, since, so,* and *therefore.*

2 Identify the Effects

- To find an effect, identify what happens.
- Clue words such as *therefore, so,* and *as a result* may fall just before the first effect.
- To find all the effects, look for more things that happen in the reading.
- These effects may be separated with commas or with clue words. Some of these words are *in addition, and, also,* and *furthermore.*

As a result, her heart and lungs have become more efficient at pumping oxygen throughout her body.

TIP **It is also possible for one effect to have several different causes.**

Effects (what happened) = Marta's heart and lungs have become more efficient at pumping oxygen throughout her body. Her leg muscles are stronger, and she has less body fat. She has been feeling happier and less stressed.

3 Look for the Cause

- Ask yourself why these events occurred.
- Clue words such as *because, caused,* and *since* may fall before a cause.

Because she jogs regularly, Marta's leg muscles are stronger, and she has less body fat.

Cause (why something happened) = Marta jogs regularly.

EXAMPLE OF **Determining Cause and Effect**

Read the following passage and think about the relationship between cause and effect.

The Bonuses of Basketball

Cause

Girls and boys who play basketball get a great aerobic workout at every practice and game. As a result, they become stronger. Ball handling strengthens upper body muscles. Running and jumping for the ball help build powerful muscles in the legs. In addition, the players burn calories while playing basketball. They run up and down the court dozens of times during a game. No wonder they get so physically fit. Of course, an improvement in heart rate and lung function benefits the rest of the body as well. Playing basketball also builds strong bones. Running and jumping on a hard court add the kind of stress to bones that helps build new cells. By the end of a game or practice, players feel strong, healthy, energized, and more confident. For these and many other reasons, playing basketball is an excellent activity for young people.

Effects

USE THIS SKILL

Determine Cause and Effect

Read the following passage. Fill in the diagram below with one cause and at least five effects.

Lu Pham has started swimming laps at a local pool three times a week. As a result, he has been feeling more energetic and alert. Instead of falling asleep at his desk each evening, he does his homework faster and more thoroughly. He sleeps more soundly at night, and he is happier and less stressed by events in his life. Also, Lu is no longer hungry all the time. Finally, Lu's swimming program is increasing his strength and endurance. His arm muscles are bigger, and he outlasts everyone in physical competitions. Lu would not give up his swimming for anything.

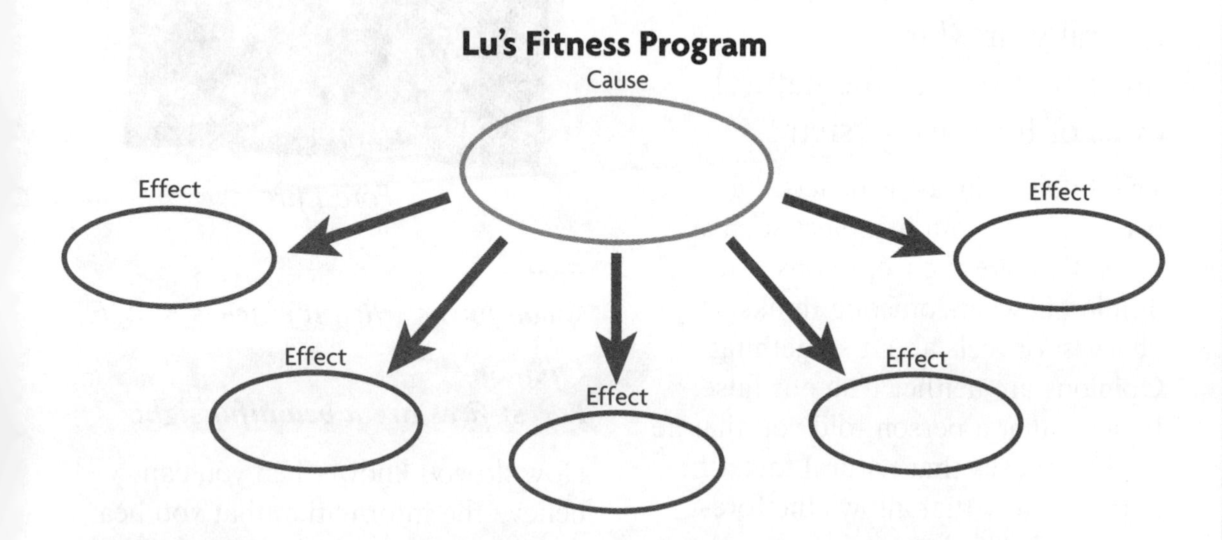

TEST TIP

You may be asked to find cause-and-effect relationships within reading passages on tests. Remember that the cause brings about the effect. The cause often comes before the effect in reading passages.

Skill 13

HOW TO

Tell Fact from Opinion

Fire: A Natural Fact

Would you believe that fire in nature is normal, even helpful at times? You may think of fire only as something that destroys living things. In fact, fire can actually improve the health of an ecosystem over a period of several years. Fires are an important part of the natural cycle of forest ecosystems.

The different ideas or beliefs that people have about a subject such as forest fires are their opinions. An **opinion** is what someone thinks, believes, or feels about something. Opinions are neither true nor false. Imagine that a person told you that he or she believes that natural forest fires burn in a way that allows the forest to renew itself. To convince you that natural fires can be good for forest ecosystems, that person might provide you with facts. A **fact** is a statement that can be proven to be true. Unlike facts, opinions cannot be proven.

Forest fire

Fact
Wood burns when it reaches 572 °F.

Opinion
Forest fires are a beautiful sight.

How do you know when you can believe the information that you hear and read? You must learn to tell the difference between facts and opinions. Facts give you the evidence you need to make informed choices. Opinions without the support of this evidence can be misleading.

STEPS IN Telling Fact from Opinion

Learn to tell facts from opinions by following these steps.

1 Find the Facts

Look for information that can be checked or proven. Scientists use facts, the evidence they learn through observation, to back up their theories or hypotheses. These are facts because they can be proven:

Fire occurs when high temperatures, oxygen, and dry fuel combine.

Lightning in Yellowstone National Park starts an average of 15 fires per year.

2 Find the Opinions

Identify statements that reflect what a person thinks, feels, or believes. Watch for clue words such as *believe, think, always, never, best, worse, more, seems, should,* and *must.*

I believe that all forest fires should be put out immediately.

Fires are so dangerous they must be the worst things that can happen to an ecosystem.

3 Check the Facts

After you have identified what you believe are the facts, make sure these facts can be proven. They shouldn't just sound true. Watch also for opinions that look like facts. Look at the statements below.

~~Fires are the number one reason plant species are endangered.~~ Not true—Researched in National Parks publication and found to be false.

~~Campers in Yellowstone National Park are careless.~~ Not always true—Some campers may be careless, but others are not. This statement could be an opinion.

Forest fires normally burn strongest in mid-afternoon. Fact—Checked in an encyclopedia and in National Parks publications.

TIP A fact can answer any of these questions: Who? What? Where? When? How many? How much?

EXAMPLE OF Telling Fact from Opinion

Read the following paragraphs from a student's report on forest fires. Think about how the facts could be proven true.

Opinion **Facts**

Fire is nature's most dramatic event. Although the actions of people start some forest fires, many fires are started by natural causes. In the Rocky Mountains, fires sometimes begin during summer thunderstorms when lightning strikes dry wood in the forest. In other places, such as California, the friction caused by an earthquake may start a fire. I think volcano eruptions are the most interesting natural causes of fires. ← **Opinion**

Facts → Natural fires have many benefits. They cut down on the amount of dead, dry wood and leaves that build up in forests so that larger, more damaging fires don't happen later. They help recycle and release nutrients into the soil. Without fires, all the most beautiful and interesting trees in the forests would die out. By leaving tree stumps behind, fires can create nesting spots for birds such as woodpeckers. Some trees, such as certain types of lodgepole pines, depend on fire to reproduce. The cones of these pines open only when they are exposed to the heat of fire. Many people who work in national parks believe that fire is nature's way of starting over.

Opinion (pointing to "Without fires, all the most beautiful and interesting trees in the forests would die out.")

Facts

USE THIS SKILL

Tell Fact from Opinion

Read the two journal entries below, which were written by a student visiting Yellowstone National Park during the fire of 1988. Find at least three facts and three opinions in his writing.

August 15, 1988

The fire in Yellowstone is getting bigger every day. The park rangers won't let my family go anywhere near the side of the park that is burning, but we can see the smoke in the air. Knowing the fire is there is scary, but it's kind of exciting too. Everyone is worried about the beautiful plants and animals that live in the park. Every day the fire burns up large patches of the sagebrush that the elk and bison eat. It's just not fair that wildflowers, birds, elk, bison, and other animals are all in danger of being destroyed in the fire.

August 16, 1988

A park ranger I met today told me that large-scale fires have been occurring naturally for thousands of years. Instead of destroying everything in the forests, these fires burn in patchwork patterns, leaving many sections less scorched and able to grow and renew the forest. Fires can even help make forest ecosystems healthier by clearing away diseased trees and nonnative plants. Now instead of trying to stop all wildfires, park officials let some run their courses. It seems like the work of a park ranger is full of challenges and dangers. I think being a park ranger is the most exciting and interesting career a person could have.

TEST TIP

Essay questions sometimes ask you to give your own opinion about a topic or issue. Be sure to write what you really think or feel. Then defend your opinion by supporting it with facts.

Skill 14

HOW TO Find the Main Idea

Lasers

It plays music, it heals, it cuts, it prints! It's the laser! **Laser** stands for "**l**ight **a**mplification by **s**timulated **e**mission of **r**adiation." Lasers are very valuable tools that are used in many different places in many different ways.

Main idea → **Supporting details**

The light from a laser beam is different from light from a flashlight or lightbulb. The beam of light from the usual sources such as lightbulbs spreads out. This is because these light waves are different wavelengths. In contrast, light from laser beams does not spread out because all of the waves in laser light travel in exactly the same direction, and all the waves are the same wavelength.

Lasers are at work in things we use every day. For example, lasers print words and graphics when you choose the print command on a computer. Because they don't scratch objects, lasers produce high-quality sound when they read the information on a music CD. Lasers are used to read bar codes at grocery and department stores, to perform surgery, and to send phone messages over fiber-optic cables. These are just a few of the ways that we use laser technology.

The paragraphs above give information about lasers. One way to better sort through information you read is to look for the main idea in each paragraph. The **main idea** is the topic or focus of a paragraph. **Supporting details** are facts or examples that give more information about the main idea. Finding the main idea can give you a better understanding of what you read and can help you remember the important parts.

STEPS IN Finding the Main Idea

Use these steps to help you find the main idea in a reading passage.

1 Skim the Reading

Study the first few sentences of a paragraph or passage. The main idea is often the first sentence in a paragraph and is called the **topic sentence.** Make sure you understand what information is given in these sentences.

2 Read Carefully

Sometimes you have to figure out what the main idea is by studying the paragraph. Carefully read the paragraph or passage. Take your time and keep reading until you understand the most important idea presented in the paragraph or passage.

Well, none of these ideas seem like a main idea. They seem like details that might support a main idea, because some are reasons and some are examples. All these details have to do with how lasers are useful to surgeons. So the main idea is probably something like: **Lasers are useful tools in different kinds of surgery.**

3 Separate Main Ideas and Details

Supporting details may be facts, reasons, or examples that develop the main idea. Generally, supporting details in a paragraph are more specific than the main idea. They may begin with or include phrases such as *for example, for instance,* or *because.* After you identify some supporting details, compare them. Ask yourself to what main idea they could all be related.

Lasers can take the place of surgical knives to cut through body tissues in some surgeries. Unlike the scalpel, the laser actually seals blood vessels as it cuts and sterilizes too. Surgeons use lasers when performing delicate surgery, such as cutting out brain tumors and repairing eye tissues. Every year thousands of people have laser eye surgery to correct their vision so they don't need eyeglasses or contact lenses anymore.

EXAMPLE OF **Finding the Main Idea**

See how one student found the main ideas of the two paragraphs below.

Main idea → The use of lasers has made all kinds of tasks faster, easier, safer, and more precise. For example, lasers have taken the place of the surgeon's knife in many medical operations. In factories, lasers are used like welders. They can melt, heat, or cut materials with great detail. Lasers are used to measure distances on land, in buildings, and even between Earth and the moon! Laser use in communications has also improved the speed that information moves through cable telephone lines. Laser printers produce good-quality, printed pages very quickly. Lasers scan bar codes, which reduces our wait time in lines at stores. Because lasers can scan a CD without scratching it, they give us a clear, high-quality sound when we listen to music on compact discs. ← **Details**

Beams of light made by lasers have been used as pointers and in light shows. ← **Main idea** Laser pointers are used to direct an audience's attention to words or images on a screen. These pointers are helpful to teachers in large classrooms and business people in meetings. Laser light can make large, colorful, moving beams that delight crowds at music concerts, stage shows, and fireworks displays. People can operate laser lights so that the movements of the laser beams are perfectly timed with music. These are just a few of the many ways that laser light can be used for both work and play. ← **Details**

Main ideas:

Lasers make doing many things safer, faster, easier, and more precise.

Laser light is used for light pointers and light shows.

USE THIS SKILL

Find the Main Idea

Read the paragraphs below. Find the main idea of each paragraph. Also find the details that support each main idea.

Just about anyone can get a laser pointer these days. Laser pointers have become very affordable because of new developments in laser technology. They are easy to find at electronics and department stores. As inexpensive as $20 or less, they are about the same price as many children's toys.

But laser pointers are not children's toys. Staring into the beam of a laser pointer can seriously injure the retinas of the eyes. Laser pointers should never be pointed directly into a person's eye. A laser light can direct such intense light energy into the eye that the effect is similar to staring directly at the sun. Damage to the retina can result in loss of sight.

In the past few years, people have become more aware of the dangers of carelessly using laser pointers. In 1997, the United Kingdom banned a certain kind of laser pointer from sale. Also in 1997, the U.S. Food and Drug Administration issued a warning about letting children use laser pointers. These actions were taken because so many people have reported eye injuries from misuse of laser pointers.

TEST TIP

On a test you may be asked to read a passage and then answer questions about it. Remember, the main idea gives information about the whole passage, not just one part of it.

Skill 15

HOW TO Take Notes

Designing Food

In the last two centuries, people's lives have changed in amazing ways. Whether in transportation, health, or housework, scientists and inventors have found ways for people to do things faster, better, and more safely. Now even the foods we eat have been improved—or have they?

When you hear about something that may affect your health or the world, you often want to learn more. What if you want to share what you learn with others? To help yourself sort through information, you can take notes on what you learn. **Taking notes** involves writing key ideas when reading or listening. Your notes will help you remember details and focus on important points. Read the passage and study the note card below it on the next page.

For many years, humans have been looking for ways to improve the plants and animals that become our food. One way that people have tried to improve food is called ***genetic engineering,*** *which involves going into the actual genes of animals and plants and transferring genetic material. Selected genes from one plant or animal are combined with the genes from another. Genes provide blueprints, or information, for living things. When genetic information from one plant or animal is put into a different plant or animal, a new set of instructions is formed. When this has been done, the plant or animal is* ***genetically altered,*** *meaning its genetic instructions have been changed.*

The notes on the card shown here were taken from the paragraph above. Pay attention to the way ideas are summarized and how short phrases are used.

Genetic Engineering

- One of the new ways people have tried to improve food
- Goes into the actual genes of plants and animals
- Combines genetic information from different plants or animals to make "genetically altered" foods

Source: Lisa Ver Hage, Genetic Engineering, SciPress, 2001, page 1

Note cards work great for taking notes, but notepaper is a good choice too. The important thing is to use a separate card or page for each kind of information. Write a heading near the top of your card or paper for each important idea.

When you take notes, keep specific questions in mind and focus on what is interesting. The next two pages describe steps you can follow for taking notes.

TIPS

- **Notes may be used later to create an outline or a first draft.**
- **Draw pictures or diagrams to help you understand difficult concepts.**
- **Note cards may be put in any order that makes it easy for you to organize ideas.**

STEPS IN Taking Notes

1 Write Questions

Before you start gathering information, think about what you really want to know. Write your questions.

1. What is genetically altered food?
2. How is it different from other food?
3. What foods are being genetically altered?
4. What are some possible benefits of genetically altered foods?

2 Gather Information

Do research and start investigating to find answers to your questions. Go to the library and use its reference sources. Look up your topic in books, magazines, newspapers, and on the Internet. Remember to use many different kinds of resources, such as videotapes, television specials, audiotapes, CDs, and written or recorded interviews with experts on your topic.

3 Organize Information

Note cards, lists, and graphic organizers are easy ways to organize information that you hear, see, or read. You can use them in different ways.

- Write the main idea at the top of each note card and write key supporting details underneath.
- Write a question at the top of each note card. When you find the answer, you can write it on the correct card.
- Cite the sources you used by writing the author, the name of the source, the publisher and copyright date, and page numbers that you used.
- Create a blank graphic organizer such as a chart or concept web. Write the main topic, then fill in the empty spaces with details as you find them.

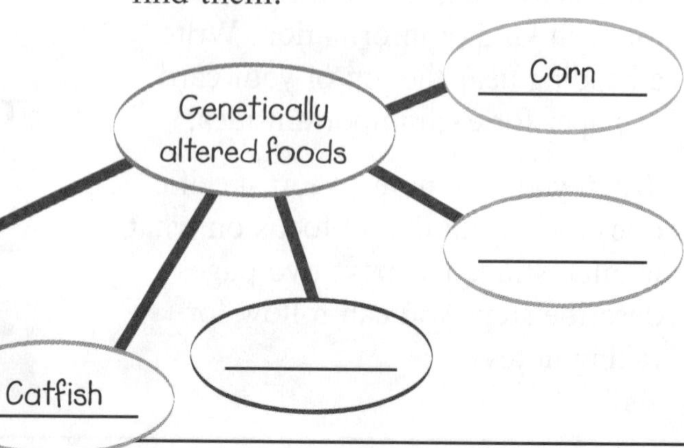

Concept web

4 Summarize and Quote

- Sometimes you will come across information in longer passages that you want to summarize. When you summarize, you condense information into just a few words. Use only the important words from the passage, just enough to help you remember what was explained.
- When taking notes, do not copy sections from your sources word for word. Rewrite what you learn, using your own words.
- If you want to include a direct quote in your notes, place quotation marks around the exact words that were said, and be sure to give the first and last names of the person who said the quote.

5 Cite Sources

In addition to citing any sources used with notes on each note card, make a list to keep track of your source information. Record the author's name, the title or the source, the publishing company's name, and the date the source was published. Write the information just as you would for a bibliography. Organize your source list in alphabetical order.

Sources

De Marco, Caroline. "The Food of the Future." Our World January, 2002.

Engineering What We Eat Videotape. Biotech Productions, 1999.

Ping, Walter. Genetic Engineering. Sun Publishing, 2001.

Magazine article

Video

Book

EXAMPLE OF **Taking Notes**

Read more about genetically engineered foods in the reference book passage below. Then study the note cards that a student created.

The Food of the Future Is Here

Don't say no to genetically engineered foods until you know what they can do. Do you know that foods such as rice can be engineered to provide nutrients such as iron and beta carotene? For millions of people around the world, genetically engineered rice may make the difference between living healthy or unhealthy lives. Some foods can be engineered to be larger in size. Everything from catfish to corn can be engineered to be larger and more plentiful so that more people may be fed. Farmers' crops increase as plants are given genes that help them resist disease, pests, and spoilage. This means that farmers do not have to use as many chemicals on their crops. Foods such as tomatoes can be given genes from flounder, a type of fish. These tomatoes will be better able to withstand cold and frost. Shots to give vaccines to children could be a thing of the past if the scientists who are working with bananas are successful. They want to provide bananas with the genetic information that will prevent some common childhood diseases. Genetically engineered foods may have many benefits for people in the future.

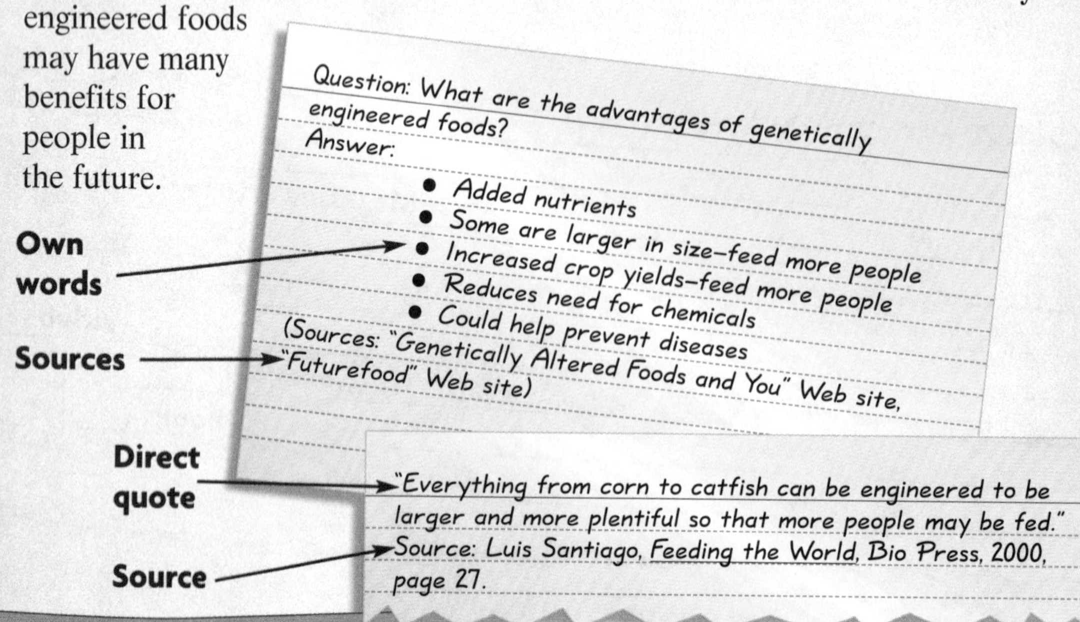

Question: What are the advantages of genetically engineered foods?
Answer:
• Added nutrients
• Some are larger in size–feed more people
• Increased crop yields–feed more people
• Reduces need for chemicals
• Could help prevent diseases
(Sources: "Genetically Altered Foods and You" Web site, "Futurefood" Web site)
Own words
Sources
Direct quote
Source
"Everything from corn to catfish can be engineered to be larger and more plentiful so that more people may be fed."
Source: Luis Santiago, Feeding the World, Bio Press, 2000, page 27.

USE THIS SKILL

Take Notes

Read and take notes on the passage below.

Stop and Think

People in favor of genetically engineered foods believe this new technology will solve many problems in the world, but have they stopped to consider the problems these foods could create? It is possible that genetically engineered plants may actually increase the use of chemicals that farmers use. These plants are sometimes sprayed with chemicals meant for nearby weeds.

Some genetically altered foods such as tomatoes contain genes from peanuts—a food that can cause severe allergic reactions in some people. Who with a peanut allergy would ever think to avoid eating a tomato? As genes from one food are added to other foods, people with allergies could be exposed to health risks.

We also have to think about what would happen if certain genetically engineered animals escaped into the natural environment. Genetically altered catfish grown on one fish farm have genes from salmon, zebra fish, and carp. They grow 60 percent faster than normal catfish. It is possible that these "super" catfish could upset the delicate balance of ecosystems if they were released into the wild.

There are many unknowns with genetically altered foods. Some people think that genetically engineered foods should not be sold and eaten until more is known about them.

TEST TIP

Two of the best ways to prepare for a test are to review your notes and get a good night's sleep. You will feel better and will be more alert when you take the test if you get enough sleep.

Skill 16

HOW TO

Predict

Nature in Danger

Have you ever observed birds at a bird feeder? Somehow they know that it's a good place to find seed. They discovered the food and learned they could get more by going back. What if the seed supply were to run out? Can you make a prediction about what would happen?

Predicting is what people do to tell what will probably happen in the future. When you make a prediction, you use your own past experiences or knowledge combined with present observations to tell what you think will happen.

Scientists who study endangered animals use predictions to determine what will happen to a certain plant or animal species. An animal or plant species becomes **endangered** when it is so rare that it is in danger of becoming extinct. When a species is **extinct,** it disappears entirely from Earth. When one species disappears, others are affected.

To make predictions about whether a species will increase or decrease in number, scientists combine what they know about how a plant or animal lives in its environment with detailed observations about present conditions.

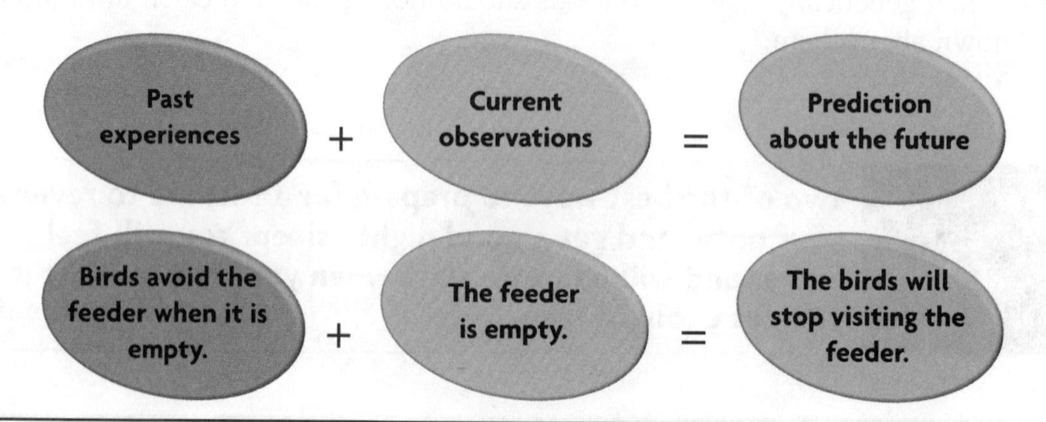
Past experiences
+
Current observations
=
Prediction about the future
Birds avoid the feeder when it is empty.
+
The feeder is empty.
=
The birds will stop visiting the feeder.

STEPS IN Predicting

Scientists make predictions when they develop hypotheses and plan investigations. Follow these steps to learn how to make a prediction.

1 Observe the Present

Pay attention to what you read and what you see going on around you. Perhaps a certain animal's environment is changing in some way. Has the stream where it lives become polluted? Are forests being cleared to make room for homes or livestock? You might want to keep notes on what you see, hear, or read about your subject.

The golden lion tamarin is endangered due to the destruction of its habitat.

2 Review Previous Experience

Ask yourself whether you know about or have experienced anything similar to the situation you are studying. For example, do you remember reading that a certain animal disappeared from an area when housing developments were built? Have you ever read that certain animals have become extinct due to overhunting?

3 Make a Prediction

Put your previous experience together with your present observations and make a statement about what you think will happen in the future. Think about the long-term effects as well as the immediate outcome. For example, do the present conditions and your past experiences and knowledge tell you that the animal or plant you are studying will become endangered or extinct?

EXAMPLE OF **Predicting**

Read this passage about tigers and one student's prediction.

Can Tigers Be Saved?

Tigers have few natural enemies, but they do have one who could cause them to disappear forever—humans. People kill tigers in order to get the high prices that their skins and other parts bring on illegal markets. Everything from tiger bones to tiger eyes are thought to cure illnesses by some people. Years ago, more than 100,000 tigers roamed across Asia. Now there are only about 5,000 left, and they are forced to live in smaller and smaller areas. The forests where they live have been cleared for firewood and other human needs. The deer and other prey that tigers feed on are disappearing because growing populations of people use this food for their own survival.

Siberian tiger

The good news is that many Asian countries now have wildlife preserves for tigers. Authorities are coming down hard on poaching and trading tiger parts. In Siberia, the number of tigers is stable, meaning their populations are neither increasing nor decreasing. Still, as human numbers increase, so do human needs. Will the pressure on tigers be so great that they cease to exist in the wild?

Present situation

- *There are only about 5,000 tigers left in Asia.*
- *More preserved areas are being set aside for tigers.*
- *Laws against tiger poaching and trading are being enforced.*
- *Tigers' food supply is disappearing.*
- *The number of tigers is stable in Siberia.*

Past knowledge

- *Large areas of tiger habitats have already been destroyed.*
- *If tiger parts continue to bring high prices, tigers will continue to be the victims of poaching and illegal trading.*

Prediction

- *I predict that some types of tigers will become extinct unless people work even harder to protect tigers and their habitats.*

USE THIS SKILL

Predict

Choose one of the endangered species listed below. Do research so you can write a description of its present situation. Read about and describe what has happened in the past to that species and other species like it. Make a prediction about whether or not you think the species will become extinct.

ENDANGERED!

- Black rhinoceros
- American ginseng (plant)
- Orinoco crocodile
- Pincushion cactus (plant)
- Nashville crayfish
- Mission blue butterfly
- Common sturgeon
- Chimpanzee
- Giant panda
- Blue whale
- Hawksbill turtle
- Red wolf
- Marine otter
- California condor
- Grevy's zebra
- Snow leopard
- Gorilla
- Golden frog
- Whooping crane
- Golden lion tamarin
- Chinese River dolphin

TEST TIP

You may be asked to make a prediction from a reading passage on a test. Look for clues in the passage that will help you decide what might happen in the future. The right answer will usually make the most sense.

Skill 17

HOW TO Make Generalizations

City Heat or Country Cool?

Are you a big-city person? Or do you prefer the quiet surroundings of the country? There may be many reasons you prefer one to the other. When comparing the two, did you consider the fact that one is usually warmer than the other?

As more and more cities expand, one thing is becoming clear to scientists. Cities, also called urban areas, are usually warmer than nearby rural, countryside areas. Scientists call warmer areas in and around cities heat islands. The idea that cities tend to be warmer than rural areas is a generalization. A **generalization** is a broad statement that is true about *most* members of a particular group. A generalization doesn't need to be true for *all* members of the group. True generalizations can be made about a group only after information is gathered about many members of the group, not just a few. Gathering information in this way is called sampling. Scientists had to compare dozens or even hundreds of urban and rural areas before they could make an acceptable generalization about them.

STEPS IN Making Generalizations

You make generalizations all the time. Use these steps to make sure that your generalizations are realistic.

1 Make Observations

Notice what goes on around you. Pay attention to what you see, hear, feel, touch, taste, and smell. Think about what you observe. When you see something happening over and over, you are seeing a pattern, or a trend. Observing trends is your basis for making generalizations.

2 Use Information You Know

Think about what you already know and how it relates to the trends you see. Use information you know from personal experiences as well as what you have learned at school and from reading books. Apply this information to what you notice to help you understand why certain things happen.

Observations:
- ✓ It feels hotter standing on the sidewalk than it does standing in the grass.
- ✓ The air in a city park feels cooler than the air along a busy city street.
- ✓ During the summer, metal on cars is sometimes too hot to touch.
- ✓ It's cooler standing in the shade of a tree than standing in sunshine.

What I Know:
- ✓ Metal and asphalt conduct heat.
- ✓ Plants are not good conductors of heat.
- ✓ Evaporation helps cool the air.
- ✓ Air pollution can trap heat.

STEPS IN Making Generalizations

3 Gather Evidence

Think about the kinds of evidence you need to gather. Plan where and how you can get this evidence. It is important to gather evidence from many members of the group you are sampling. When gathering evidence, use numbers to tell about your sample size, such as "The average daily temperatures of 20 cities were recorded each day for 90 days."

Kind of evidence:
Downtown air temperatures for 30 cities and their nearby rural areas

Where to get evidence:
Check local weather stations, National Weather Service, and Internet.

Temperatures (°F) in and around Cities

	January 2		January 15		February 5	
	City	Rural	City	Rural	City	Rural
New York	20	15	5	1	21	17
Atlanta	31	29	39	37	44	42
Chicago	2	-5	24	20	33	30
Dallas	35	31	42	39	55	52

4 Analyze Evidence

When you have gathered your evidence, think about what it shows. Does it support the trend you thought you observed? If a trend is supported in the samples you have analyzed, you are ready to make a generalization.

Words commonly used when making generalizations:

- ✓ usually
- ✓ mostly
- ✓ frequently
- ✓ generally
- ✓ often
- ✓ typically
- ✓ tends to

5 Make a Generalization

Combine your evidence with your experiences and other information you know to make a generalization about a particular subject or group. Remember that a generalization is true of *most* members of the group; it does not have to be true of *all* members of the group.

Below is a generalization that can be made by gathering and analyzing the evidence on the previous page.

Air temperatures in cities tend to be higher than air temperatures in surrounding rural areas.

EXAMPLE OF Making Generalizations

Read how one student used evidence and observations to make a generalization about temperatures in cities and rural areas.

- Kyle lives on a farm in a rural area outside of a large city. On one of his many trips to the city, Kyle noticed the following:

Observations

- Snow quickly melts on city streets, but the country roads are covered with snow for days.
- In the summer, there is usually a breeze in the country and the air feels cooler than in the city.
- There are more cars in the city than in the country. The air around moving cars is very warm.
- There are more trees and plants in the country than in the city.

- Kyle knew there were reasons for these differences. For example, he learned in school that some materials, such as concrete and metal, absorb and conduct heat.
- Kyle watched the local news for several weeks and recorded the high and low temperatures of both the city and his rural area. He decided he needed to keep records regularly over the coming months to be sure the sample size was large enough.
- Once Kyle had gathered all his evidence, he analyzed the data he collected.

Generalization

Temperatures in the city are usually higher than temperatures in surrounding rural areas.

USE THIS SKILL

Make Generalizations

Read the text below and answer the questions.

In the large city where Rosa lives there are several different areas. One area has mostly factories and tall office buildings. Another area has neighborhoods filled with older homes, lawns, and trees.

Rosa knew that neighborhoods are usually cooler than areas with factories and tall buildings. She decided to compare the temperatures inside her city for several months. She asked friends living in these two areas to record the temperature at exactly 4:00 P.M. each day for four months.

The chart below shows the evidence that Rosa gathered.

Area	Average Temperatures (°F)			
	June	July	August	September
Neighborhood	84	84	86	83
Factories, Offices	86	87	89	85

1. What information did Rosa know?
2. What evidence did Rosa gather?
3. What generalizations can you make about the temperatures in Rosa's city?

TEST TIP **When asked to make a generalization on a test, be sure your generalization is based on information provided, not on your own opinion.**

Skill 18

HOW TO

Make a Decision

How Does Your Garden Grow?

Imagine that you are given a small piece of land with the opportunity to create your very own garden. How would you decide which seeds to buy and plant in your garden?

Vegetable garden

There are lots of different types of gardens. A person who wants to enjoy the beauty of flowers might want to plant a flower garden. Some gardeners may choose to plant a fruit or vegetable garden. They may be most interested in the food that a garden can supply. Other people might want to plant a combination of fruits, vegetables, and flowers.

A garden, just as any other project, requires a plan. When you put together a plan, you have to make decisions. **Making decisions** involves looking at different options and making choices that will help you reach a goal.

After you make one decision, you often have additional decisions to make. For example, if you choose to grow a fruit and vegetable garden, you might have to decide how big you want the garden to be, what types of fruits and vegetables you like to eat, which are suited to the conditions in your yard, and how you plan to take care of the plants. To make any kind of decision, follow the steps on the next page.

STEPS IN Making a Decision

1 Set Goals

Think about what is important to you. What do you want to accomplish or have happen? *Do you want to grow pumpkins that can be harvested in the fall? Do you want to grow sweet corn that your family can enjoy in August? Would you like to have flowers blooming in springtime?*

2 Identify Options

Look at all the possible choices. What are some different ways you can meet your goals? Make a list of all your options. You can use this list to help you brainstorm ideas for ways to meet your goals.

3 Consider Your Options

Consider which options will help you meet your goals. Think about what else you might need to know about your choices. Is every option really possible? Learn as much as you can about your choices to discover their advantages and disadvantages. Reread your list of options and carefully consider each one.

Does a certain type of fruit or vegetable grow well in the climate where I live? How will it do in the soil that I have? Does it need lots of sun? Does it need lots of water? Which vegetables are my family's favorites?

List the good points and bad points of each option. Think about your own beliefs and how each choice will affect other people.

4 Make Your Decision

Think about the immediate and long-term consequences or outcomes of each choice. You can also rank the options from best to worst, according to how well they meet your goals. Make your final decision.

5 Evaluate Your Decision

Ask yourself whether your decision is working and whether it feels right. Keep an open mind to new information. Reconsider your decision if your choice does not help you accomplish your goal. *The tomatoes I planted do not look healthy. I think the soil in my garden is too sandy. Next year I will plant carrots, which should do well in sandy soil.*

EXAMPLE OF Making a Decision

See how young people at the Albany Recreation Center used a comparison chart to help them decide which three kinds of foods to plant in their garden.

The young people at Albany Recreation Center thought about what they like to eat and what might be fun to grow. To narrow their choices, they compared the needs of each plant to what they could provide for it. After considering several options, they decided which three they would plant in their garden. ← **Set goals**

Identify Options → Food

Consider options → Soil Needs, Light Needs, Advantages, Disadvantages

Food	Soil Needs	Light Needs	Advantages	Disadvantages
Corn	Rich soil—full of nitrogen	Full sun	Vertical growth, requires little space	Too much shade for nearby vegetables
Zucchini	Fertile, organic matter, able to retain moisture	Full sun	Easy to grow	Needs extra space
Carrots	Deep, loose, sandy loam	Sun/some shade OK	Will use little space and can be planted early	Require thinning
Tomatoes	Loose, rich with organic matter	Full sun	Will taste much better than store-bought	Must buy plants, not seed
Asparagus	Sandy, slightly acidic loam	Full sun	Will grow back year after year	Must wait three years before harvesting

Our decision: zucchini, carrots, tomatoes ← **Make a decision**

USE THIS SKILL

Make a Decision

Study the garden description below and read the plant descriptions. Make a decision on two or three plants you would like to add to this garden. Explain why you made this decision.

This large garden has dark, rich, slightly acidic soil. The soil is moist but well drained. Corn is planted at one end of the garden, providing a large area of shade. The rest of the garden gets full sun throughout the day. There is enough room in the garden for three plants with medium-sized space requirements, or two plants with large space requirements.

Beans *grow in neutral to slightly acidic soil that is rich in organic matter. Beans grow 4 inches apart in full sun. Bean plants may shade other crops.*

Lettuce *does well in shade cast by taller plants. It prefers rich soil, grows quickly, and rarely gets diseases. Lettuce should be kept moist at all times. Allow plenty of space in between seeds, and thin plants as they develop.*

Basil *is an herb that thrives in full sun. Plant it in rich, loose soil that is not acidic.*

Chives *are a hardy herb that can grow in full sunlight or part shade. Chives need moist, well-drained soil. They usually last through the winter.*

Strawberries *need a sunny, warm location and a very large space in which to grow. They grow best in well-drained, fertile soil and require frequent weeding. Strawberries can spread diseases to other plants. They produce luscious fruits year after year.*

Radishes *take up little space and grow very quickly. They prefer sun but will tolerate some shade. Radishes like sandy, moist soil covered by mulch.*

Potatoes *do well in full sun and prefer slightly acidic soil that is well drained. Potatoes can carry diseases and attract pests. They use a medium-sized space.*

TEST TIP

Some tests may ask you to choose between two or more options or results. Make sure you read all the answer choices carefully before you make a decision.

Skill 19

HOW TO

Work in a Group

Building to Last

When architects design a building or other structure, they have to keep the structure's purpose in mind. Most importantly, they want to design structures that are safe, that are structurally sound, and that will last.

Architects are professional people who plan, design, and direct the building of structures. Architects make many decisions, including decisions about what shapes and materials to use. They base these decisions on what forces will be acting on the structure as well as what weight the structure will need to support. One of the most stable and rigid shapes used in building things today is the triangle. Other common shapes used in the building of structures are rectangles and arches.

The TransAmerica building in San Francisco

Architects work with a group of people to create the final product. Learning how to work with others in a group to achieve a common goal is an important skill. Suppose you are going to work in a group to build a structure. You will need to make a lot of decisions to complete your project. Follow the steps on the next page to learn how to work in a group.

STEPS IN **Working in a Group**

1 Define Your Goal

The first thing your group needs to do is decide on a goal. Ask yourselves exactly what you want to accomplish. Discuss options and ideas, making sure each member of the group shares his or her opinion. It is important to listen carefully to what each group member has to say.

Goal: *We want to build a structure that will add something interesting to our classroom.*

Questions: *How will we use the structure? How will we get materials? Who can help us design and build it? How long will this project take?*

Collect information to help narrow and define your goal. When you have made a group decision about your goal, write it down.

2 List and Choose Tasks

Talk about what tasks, or jobs, need to be done to complete the project. As a group, create a list of everything that needs to be done, as well as all the materials that will be needed. When you are finished, have group members choose one or more tasks. Try to divide the tasks so that all members have about the same amount of work.

3 Make and Keep a Schedule

Create a schedule that lists the tasks and the dates they should be completed. As tasks are done, cross them off or mark them in another way to show the progress you are making.

<u>Group Project</u>
Goal: ______
What needs to be done: ______
Materials we will need: ______
Tasks for each group member: ______
Schedule: ______
Task: ______
Date due: ______
Task: ______
Date due: ______
Task: ______
Date due: ______

EXAMPLE OF Working in a Group

Read below to see how four students made a group plan to accomplish a goal.

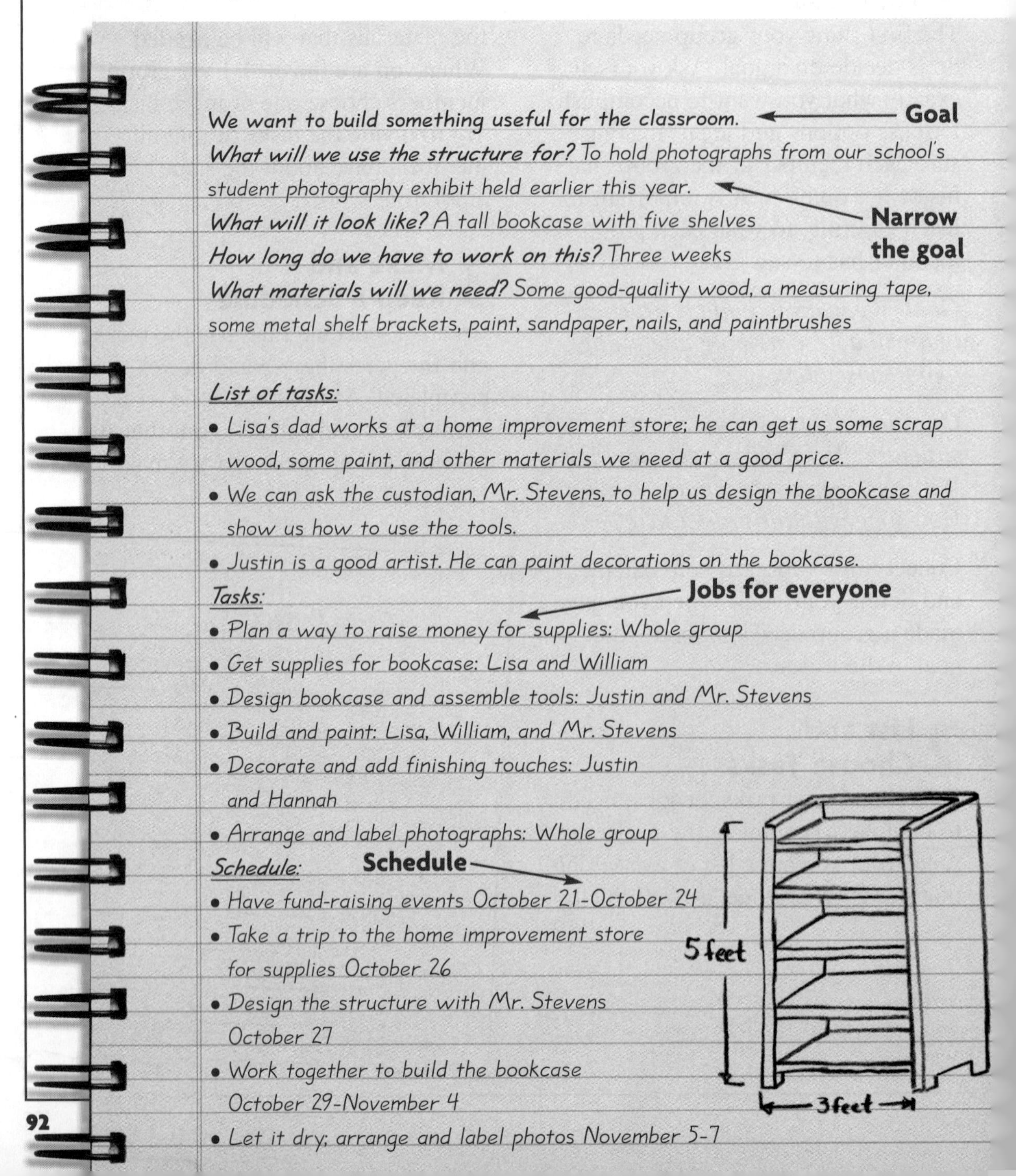

We want to build something useful for the classroom.

What will we use the structure for? To hold photographs from our school's student photography exhibit held earlier this year.

What will it look like? A tall bookcase with five shelves

How long do we have to work on this? Three weeks

What materials will we need? Some good-quality wood, a measuring tape, some metal shelf brackets, paint, sandpaper, nails, and paintbrushes

List of tasks:

- Lisa's dad works at a home improvement store; he can get us some scrap wood, some paint, and other materials we need at a good price.
- We can ask the custodian, Mr. Stevens, to help us design the bookcase and show us how to use the tools.
- Justin is a good artist. He can paint decorations on the bookcase.

Tasks:

- Plan a way to raise money for supplies: Whole group
- Get supplies for bookcase: Lisa and William
- Design bookcase and assemble tools: Justin and Mr. Stevens
- Build and paint: Lisa, William, and Mr. Stevens
- Decorate and add finishing touches: Justin and Hannah
- Arrange and label photographs: Whole group

Schedule:

- Have fund-raising events October 21-October 24
- Take a trip to the home improvement store for supplies October 26
- Design the structure with Mr. Stevens October 27
- Work together to build the bookcase October 29-November 4
- Let it dry; arrange and label photos November 5-7

USE THIS SKILL

Work in a Group

Read the passage below, then work with other students to develop a group plan.

Imagine that your class wants to build a structure for your school library. Think about what kind of structure might be useful, such as a display rack, bulletin board, or shelves. Also consider the materials and other resources you have available. Keep in mind that the structure should be sturdy and safe. Working with a group of classmates, come up with a plan you can follow to reach your goal.

TEST TIP

You should have a plan when you take tests. For example, you should first scan the test to see how long it is and what types of questions there are. Answer the questions that you are sure of first, then go back and deal with the more difficult questions with the time you have left. Be sure you don't skip any pages by mistake.

Writing and Research Skills

Skill 20

HOW TO

Make a Learning Log

Fascinating Snakes

Does the thought of snakes send shivers up your spine, or do you find snakes fascinating? Movies, stories, and television shows make them seem dangerous and deadly, but are you aware that only about one third of all snakes are actually poisonous?

You may already know that a snake is a reptile. **Reptiles** are cold-blooded animals that have backbones. Snakes, like other reptiles, have scaly, dry, waterproof skin. Their thick skin and scales protect their bodies from injury and from drying out. Snakes do not have ear openings or eyelids. Snakes move close to the ground, as do other reptiles. Snakes, however, move on their bellies while other reptiles move on short legs. Snakes eat insects and other animals. Some move quickly and grab at their prey with their mouths, while others wrap their bodies around their prey and constrict, or squeeze, the prey. A few species use sharp fangs to inject their prey with poisonous venom.

Ribbon snake

One way to learn more about an animal such as a snake is to keep a learning log. A **learning log** is a notebook that contains blank pages for recording information about something you are studying. It is different from a regular journal, which is used for writing ideas and information about personal experiences. Use a learning log as a tool to focus on a subject you want to explore.

Sometimes an exploration will involve asking questions and doing research in a library or on the Internet. Other times it will involve doing an investigation to answer a question or test a hypothesis. When scientists use learning logs or "science diaries," they record exactly how they set up their investigations, what they observe, and what they discover or conclude. They record questions and ideas they think of as they work.

You can use a learning log to record notes about resources you want to use, to create drawings and diagrams, to record observations, and to write the definitions of new words or science terms.

The learning log entries below show one student's explorations on the subject of snakes.

March 16

Snakes

Today our science class went to the Natural Sciences Museum. I got to hold a blacksnake, and it felt very dry and smooth. One of the other snakes at the museum just had babies and its leathery looking eggs were still lying around its terrarium. The mother snake spends a lot of time flicking her tongue in and out. My friend Cameron told me that snakes use their tongues to smell.

Note: *Go to library on Tuesday and find out how snakes use their tongues to smell.*

March 17

A Complete Reference Guide to Reptiles and Amphibians,

by Beth Briggs, SciPress, 2000, page 178.

"Snakes use their tongues to flick scent particles from the air or a nearby surface back to a special organ in their mouths called the Jacobson's organ. This organ consists of two hollow sacs with nerve endings that are sensitive to odors. When following its prey, a snake uses its tongue to pick up scent and send it back to the Jacobson's organ."

STEPS IN Making a Learning Log

Follow these steps to make your own learning log.

1 Include Questions

Write questions you have about your subject, and note the person or resource you plan to use to answer these questions. In some cases, a question may inspire you to begin a new investigation.

Question: How do snakes move?
(Find answer by: 1) looking in books at the public library; 2) Natural Sciences Museum exhibit; 3) doing research on the Internet.)
What I found out:
There are four different ways:
• Flex muscles and form loops that push against surfaces to move the snake sideways
• Contract muscles and pull the scales of its belly forward to move in a straight line
• Coil body and push forward
• Lift the main part of its body off the ground and sideways, using its head and tail as supports ("Sidewinding"). After the body lifts off, the head and tail follow.

2 Choose an Approach

Decide *how* you want to learn about your subject. For example, if your subject is an animal that lives in a faraway place, you may decide to conduct research on the Internet. If your subject is something that can be observed firsthand, such as the way a plant or pet in your house grows or behaves, you can set up an investigation. In the learning log, you will record the details of your firsthand observations and/or notes from your research.

3 Organize Pages

If you are studying more than one subject, you may wish to keep a separate notebook for each subject. You could also keep one notebook with separate sections for each subject.

4 Record Your Information

- If you do library or Internet research, sum up useful information and write it in your own words, unless you plan to quote it later. Be sure to cite any sources you use.
- When doing science investigations, it is important to record the exact times and dates of observations. Use your senses of smell, sight, sound, touch, and even taste, when appropriate, and write descriptions telling what you have observed by using these senses.
- Use pictures or diagrams when recording information. Create your own drawings to help you understand a description or remember something you observed firsthand.

Learning Log: Grade 5 Science

Watching a Corn Snake Shed Its Skin

Day 1

9:00 A.M.: The snake's eyes look like they are clouding over. The snake's skin looks dull and slightly dusty.

Day 2

1:00 P.M.: The snake has buried itself under the moss and wood chips in its tank.

Day 3

8:00 P.M.: The snake remained hidden the entire day.

Day 4

9:00 A.M.: The snake slowly crawls out from under the moss and wood chips. 2:00 P.M.: A few hours later, I see the snake begin crawling out of its skin! The old skin looks like it is being turned inside out. The new skin underneath looks shinier and brighter in color.

Day 5

10:00 A.M.: The snake has crawled nearly all the way out of its old skin. The snake's eyes are clear and bright. The old skin is white and scaly-looking. 3:00 P.M.: The old skin has come all the way off. The snake is moving around the tank and flicking its tongue in and out.

EXAMPLE OF **Making a Learning Log**

During their unit on reptiles, the students in Ms. McCabe's fifth-grade class observed their classroom pet, a garter snake. Read the learning log kept by one student. Note the details and specific times recorded.

Rachael Stein
Science–Ms. McCabe

Learning Log

Question

Question: Because snakes are cold-blooded animals, how does the garter snake regulate its body temperature?

How does our class's garter snake react to changes in temperature in its large terrarium tank? A heating pad heats the bottom of the tank at night. In the morning the pad is unplugged. Above the heating pad is a layer of bedding made of tree moss and sawdust.

Approach

We will place the tank near a window where it will get direct sunlight for four hours on the day of the investigation. The snake has a water bowl, hollow log, and several small tree branches in its tank. We will use a thermometer to measure temperature changes.

Record information

Organize

Wednesday, February 7

8:30 A.M. I arrive at school and find the snake coiled on the bedding. The temperature at the bottom of the terrarium reads 79°F. Ms. McCabe unplugs the heating pad.

10:30 A.M. The snake climbs up the branches near the sunlight where it is warmer. The temperature near the base of the terrarium is 68°F.

12:30 P.M. The snake crawls up and down the branches. Sunlight appears to have warmed up the entire terrarium. The thermometer reads 80°F.

2:30 P.M. Direct sunlight hits the terrarium. The temperature reads 95°F. The snake is coiled up under the hollow log, away from the sunlight.

Answer to our question: The snake seeks warmth but avoids extreme heat. The garter snake regulates its body temperature by seeking areas in the tank that are neither too warm nor too cold.

USE THIS SKILL

Make a Learning Log

Using one of the sample questions listed below, create your own learning log in a blank notebook. Record your questions, ideas, notes, and observations in the learning log for one week.

Learning Log Topic Ideas	
• How does the weather in my neighborhood change?	• What kinds of plants and animals live in a local ecosystem (such as a river, pond, park, or woods)?
• What does our class pet (or a pet at home) do each day?	• What are some questions I have about what we are studying in science class?
• What are the daily activities of fish in an aquarium?	• What are the definitions of new words that I hear at school or find in books?
• What happens each day at a bird feeder?	• What are the definitions of science terms and units of measurements I have heard but don't understand?
• What changes and activities are there in a garden?	

TEST TIP

On a test you may be asked to support a conclusion. Be sure you know if you are supposed to support the conclusion with facts from a passage or your own opinions and experiences.

Skill 21

HOW TO
Write an Outline

Astronomy of Ancient Peoples

How do you feel when you gaze into the clear night sky? What do you think when you see a sky filled with stars and a bright moon? One thing that hasn't changed over thousands of years is that people are still fascinated by the universe.

Astronomy is the study of objects in the sky, such as the moon, stars, and planets. Many of the peoples who lived long ago studied the sky to help them tell time and directions, to know when to plant and harvest crops, and even to predict events such as eclipses. The study of how ancient cultures used the science of astronomy is called **archeoastronomy.**

As you learn about new topics such as archeoastronomy, you may find it helpful to organize what you read and hear into an outline. An **outline** is a formal way to organize ideas or notes. In an outline, you organize information from general ideas to specific ideas. Outlines are useful in planning a writing that includes a large amount of information about a subject.

Ancient peoples may have built Stonehenge as a kind of calendar using the position of the sun.

STEPS IN Writing an Outline

1 Write a Title for Your Outline

Decide on a short title that tells the subject of your outline. Write the title at the top of the page.

Astronomy in Ancient Times

2 Identify Main Topics

Include at least two main ideas about the topic to be used as headings in your outline. Number each heading with a Roman numeral. Capitalize the first word of each heading.

I. People in many ancient cultures built structures for studying the sky.

3 Identify Subtopics

Write details about the main ideas under the headings. Details written in short phrases or sentences on outlines are called subtopics, or subheads. Indent and use capital letters before each subtopic. Each main topic heading should include at least two subtopics.

Astronomy in Ancient Times

I. People in many ancient cultures built structures for studying the sky.

A. Mayans built astronomy observatories.

B. Egyptians aligned their pyramids with stars.

C. Ancient peoples in England may have built Stonehenge to help study movements of the sun, moon, and stars.

4 Review Your Outline

Reread your outline to see if all the important information about your topic is under the correct main heading or subtopic.

✔ Does your outline have two or more main headings?

✔ Does every main heading have at least two subtopics?

✔ Is the outline focused on the topic?

EXAMPLE OF Writing an Outline

Read the following outline of an article about archeoastronomy. Notice the structure of the outline and how numbers and capital letters are used.

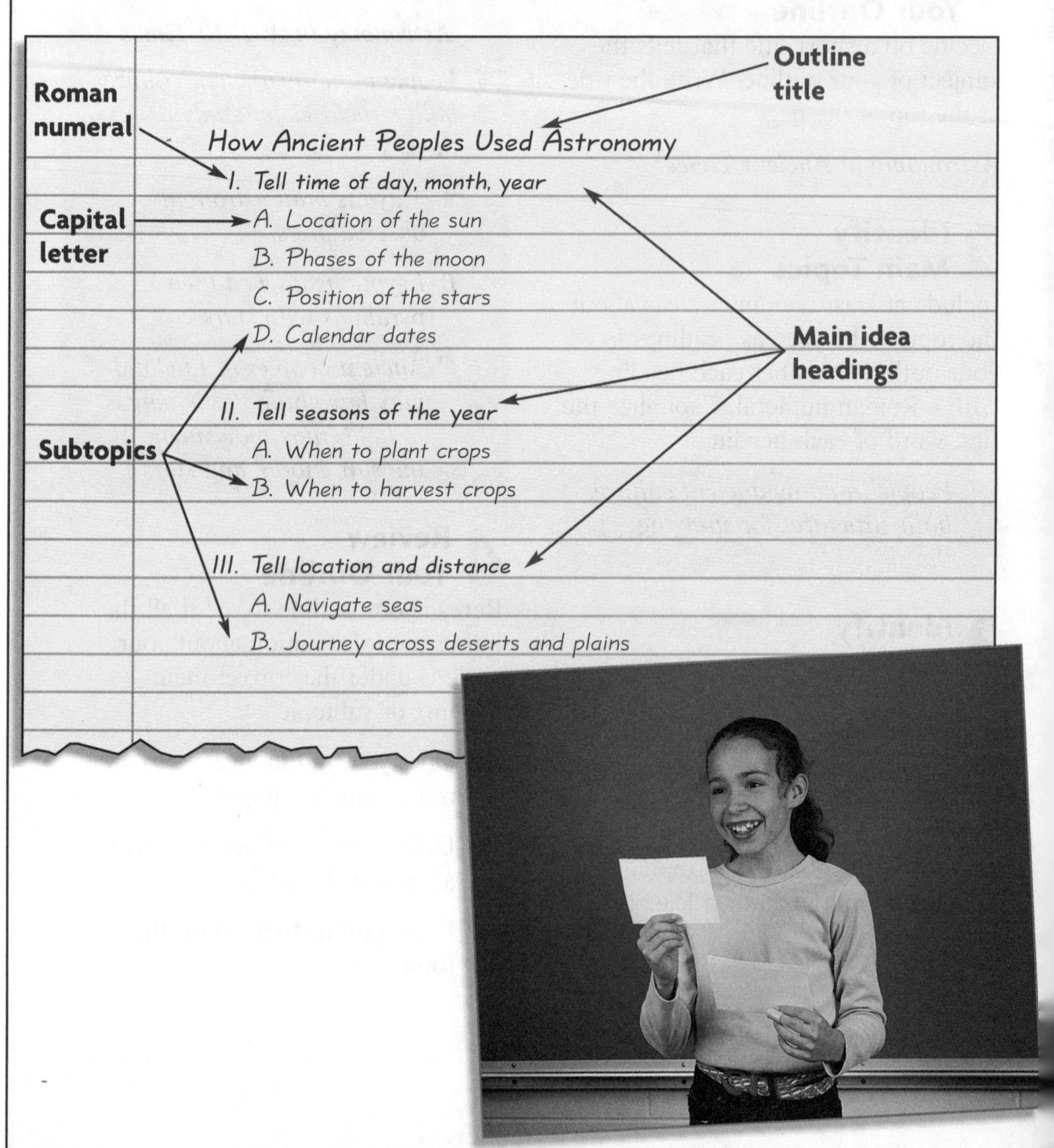

Roman numeral
Outline title
How Ancient Peoples Used Astronomy
I. Tell time of day, month, year
Capital letter
A. Location of the sun
B. Phases of the moon
C. Position of the stars
D. Calendar dates
Main idea headings
II. Tell seasons of the year
Subtopics
A. When to plant crops
B. When to harvest crops
III. Tell location and distance
A. Navigate seas
B. Journey across deserts and plains

USE THIS SKILL

Write an Outline

Read the information below on ancient calendars. Then think about the questions that follow. Use this information to write an outline on this topic.

Ancient Calendars

Many ancient civilizations used their study of the sky to develop calendars. Most of the ancient calendars were based on the positions of the sun (solar), the moon (lunar), or the stars. Some of the oldest calendars are the Egyptian and the Chinese calendars. The Egyptian calendar was a solar calendar that had 365 days. The Chinese calendar was a lunar calendar of twelve periods of 29 or 30 days. The Hindu calendar is based on a solar year, and it has 12 months and 365 days. It is the calendar of present-day India. The Muslim calendar is a lunar, 12-month calendar that has 29 or 30 days in each month. This calendar does not relate to the seasons.

The Roman calendar, taken from the Greek calendar, had ten months based on the solar year. This calendar was eventually replaced with a lunar calendar that had 355 days. Julius Caesar, leader during the late republic of Ancient Rome, wanted to have one calendar for the entire Roman Empire. He had a solar calendar developed with 365 days. This is the origin of the calendar the Western world uses today.

1. What is the main idea of this reading passage?
2. What would be the two main headings?
3. What would be the subheads under each of these main headings?

TEST TIP

Some tests may ask you to organize items into main topics and subtopics. Use what you have learned about writing outlines to help you decide which items are main ideas and which give more detailed information about main ideas.

Skill 22

HOW TO

Write a Summary

Saving the Sun's Heat

Like the wind, the sun is one source of energy that is always with us. Did you know that we can capture the sun's energy, called **solar energy,** using solar collectors? Solar energy can be used to heat our homes and to generate electricity. However, there is one major problem with the collection of solar energy. Solar collectors may not be able to meet our energy demands on cloudy days and during the night.

Solar energy collector

There is a lot to learn and remember about solar energy. One way to remember what you read is to write a summary. A **summary** is a short passage that tells the most important ideas from a piece of writing. When you write a summary, you arrange main ideas from a passage in a way that makes sense and is easy to read.

A summary is always much shorter than the original piece of writing. This is because it doesn't include a lot of details. Details add information, often in the form of examples, facts, and interesting descriptions. If you can't decide if a piece of information is a main idea, ask yourself if the piece of writing still makes sense without it. If it doesn't, chances are it's a main idea. Look at the chart on the next page to see some main ideas and details about solar energy.

STEPS IN Writing a Summary

Follow these steps to learn how to write a summary.

1 Read Carefully

Start by quickly reading the passage—don't slow down to examine details. After you read, ask yourself what the passage was about. You should be able to remember some of the main ideas. Next read the passage more carefully to find details. Understanding the supporting details will help you better understand the main ideas.

2 Take Notes

As you read, look for topic sentences. They often come at the beginning or end of a paragraph and contain the main idea. Take notes on the main idea of each paragraph. Don't copy word for word. Be sure to use your own words.

Solar Energy	
Main Ideas	Details
Solar batteries allow us to store solar energy for later use.	Solar batteries are used in spacecraft as a source of electricity to operate equipment aboard.
Made from a special material called silicon, solar cells are used to change energy from sunlight into electricity.	The silicon is cut into flat, round pieces about the size of dinner plates.
Passive solar heating is the direct use of the sun's energy for heat.	The Romans may have used passive solar heating centuries ago.

3 Organize and Write

Decide how to organize the main ideas you noted. Study your note cards to see if there is one central idea that you can use as the topic sentence for your summary. Starting with this topic sentence, write a first draft of your summary. Include only important information, and write complete sentences.

4 Revise, Edit, and Publish

Compare your summary to the original passage. Make sure that you have used your own words. Add important ideas that you may have left out. Check and correct mistakes in spelling, punctuation, and grammar. Write a final, neat draft of your summary on a clean sheet of paper.

EXAMPLE OF **Writing a Summary**

Read the following passage and the example summary below.

Solar Collectors

Main idea

Supporting details

Main idea

Solar collectors take in energy from the sun's rays. Solar collectors come in all shapes and sizes. They may be boxes, frames, or even rooms.

Solar collectors need four main parts in order to work. On the outside of a solar collector are clear covers that let in the sun's energy and then trap the heat it produces. The cover is often made of glass. On the inside, dark surfaces called absorbers soak up the light energy and change it to heat energy. The absorbers may be metal sheets or containers filled with water, rocks, or bricks that are dark or have been painted a dark color. These materials can store the heat energy. The third part of solar collectors is insulation. Insulation materials stop the heat that is trapped in the solar collector from escaping. Insulation is made of materials that are not good conductors of heat, such as cork, felt, or cotton. Finally a solar collector moves the heat to the place where it is needed. For example, vents, ducts, fans, pipes, and pumps carry heated air or water from the solar collector to the rooms of a house or other building.

Supporting details

Summary

Solar collectors collect energy from the sun. Solar collectors have four main parts: (#1) Clear covers that let in and trap solar energy; (#2) Absorbers made of dark surfaces that soak up heat; (#3) Insulation that keeps the heat from getting away; and (#4) Vents, fans, or pipes that take the heated air or liquid to the place it is needed.

USE THIS SKILL

Write a Summary

Write a summary of the following written passage.

Energy Source Choices

There are many problems with the traditional energy sources we still use to heat our homes. These sources include the fossil fuels of natural gas, oil, and coal. Fossil fuels are energy sources that are formed from the decayed remains of animals and plants over a long period of time. Fossil fuels are nonrenewable, which means their supply will probably run out some day. Gas, oil, and coal also cost money. In addition, the use of these kinds of sources causes problems for the environment. For example, gasoline used in cars adds smoke, carbon monoxide, and other chemicals to the air we breathe.

Because of the problems with fossil fuel energy sources, the use of solar energy for heating is becoming more and more important. The sun's energy is a renewable, free resource. Solar collectors are fairly inexpensive and easy to build. If more people put solar heating systems in their homes, then less fossil fuels would be used, and the environment would be cleaner.

TEST TIP

Some tests may ask you to write a summary of a passage. Keep in mind that a summary can be very short. Write about the most important ideas in the passage in your own words.

Skill 23

HOW TO

Write a Description

Wonders of the Rain Forest

Imagine living in a tall building with many floors. Each floor provides a different type of area. For example, one floor might have apartments. Another floor might have a gym, and there might be restaurants and stores on other floors. Like tall buildings, rain forests are full of activity at every level. In order to survive, plants and animals at each level depend on what that level provides.

Tropical rain forests are forests in areas that get huge amounts of rain and have daytime temperatures that stay close to 80 °F. Some rain forests get more than two feet of rain in one month—more than the wettest cities in the United States get in a whole year! These very warm and wet climates create rich environments that are home to more than half the world's plant and animal species. New species are discovered every year in rain forests. The diverse plant life in rain forests also provides people with foods, spices, valuable medicines, and wood for building materials and fuel.

The richness of the rain forest goes beyond its tallest trees and smallest insects. It extends to the very air we breathe. Some people call rain forests the "Lungs of the Earth" because their miles of dense plant life take in carbon dioxide and release huge amounts of oxygen into Earth's atmosphere.

How do you begin to describe a place as rich and varied as a rain forest? It helps to organize your description. Descriptions may be organized as if you are looking at an object or a place from top to bottom, from the left side to the right, from far away to close-up, or any other way that makes sense for what is being described.

The following description provides an overview of the rain forest. It goes from top to bottom. Notice how the writer uses **location words** such as *above, below,* and *beneath* to guide the reader through the description.

At the very top of the rain forest is the emergent layer, formed by the tops of the very tallest trees poking through the mass of other trees. High-flying birds such as the harpy eagle perch in and soar above the emergent layer watching for prey—the sloths and large monkeys that live below in the canopy. The canopy gets its name because of the way its tall trees spread their branches and form a continuous layer, almost like an umbrella. The canopy shades the layer beneath it called the understory, where shorter trees and shrubs grow and where reptiles, insects, and vines thrive. At the bottom of the understory is the dry forest floor where peccaries, ants, and other floor-dwelling animals live.

South American rain forest

STEPS IN Writing a Description

Follow these steps to learn how to write a description.

1 Gather Details

- Collect as many details as possible. Do research or record details from your own observations, if possible. Include measurements and other scientific details when needed.

From nose to tip of tail, the green iguana measured 51 inches.

- Good descriptions appeal to the reader's sense of sight, touch, sound, smell, and taste. Using details about the five senses can make a description more lively and interesting. Think about what words will describe the properties of what you are observing.

The plump, smooth, rain forest cacao bean surprised me with its bitterness.

2 Organize Your Description

Making a graphic organizer is a useful way to record and organize details. Charts, lists, concept webs, spider maps, and network trees are some of the many types of graphic organizers you can choose to make.

Exploring a Rain Forest River	
Sight	• Sunlight sparkling on dark water • Crocodiles resting on muddy, brown islands • Brilliant red, yellow, and green birds perched in the trees
Smell	• Musty river water smell • Heavy, rich smell of damp earth and plants • Sharp, woody smoke from nearby slash and burn fire
Sound	• Quiet, gurgling water • Chattering birds • Oars bumping the boat • Buzzing insects
Taste	• Lukewarm water from my water bottle
Touch	• Warm river water • Hot sun

Organize your description in a way that helps readers follow along and clearly imagine your topic. Try using one of these three approaches to organize your description:

- **Top to bottom (or bottom to top):** Begin with what is at the top and move to the bottom. This approach best describes tall things such as towers, volcanoes, and trees.
- **Left to right (or right to left):** Start with what is on the left and move to the right. This approach works well for horizontal things such as rivers, cities, and forests.
- **Near to far (or far to near):** Begin with what is closest to you and continue with what is in the background. This is a good way to describe scenes such as landscapes, crowds of people, and large spaces.

3 Create a Draft

When you have collected your details and decided how to organize your description, you can begin writing. Begin with a topic sentence to tell the reader what you are about to describe. Fill the paragraph with vivid details. End with a closing sentence to let your reader know you have finished.

TIPS

- **Include colorful adjectives such as sticky, scaly, and piercing.**
- **Use action verbs such as flutter, slip, and howl.**

Topic sentence

The spider monkey seemed to be staring at us. Its small, furry head framed its glittering, dark eyes. Its long, dark fingers grasped the tree branch, and in one smooth movement, it swung hand over hand to the next tree, using its snake-like tail as an extra support. . . .

Closing sentence

After entertaining us for several minutes, the acrobatic little monkey flipped and glided away through the breezy, swaying treetops.

4 Revise, Edit, and Publish

Revise and edit your first draft by checking your choice of words and making corrections to spelling and grammar. Write a neat, final copy on a fresh sheet of paper.

EXAMPLE OF **Writing a Description**

Read the following description. Pay attention to how it is organized.

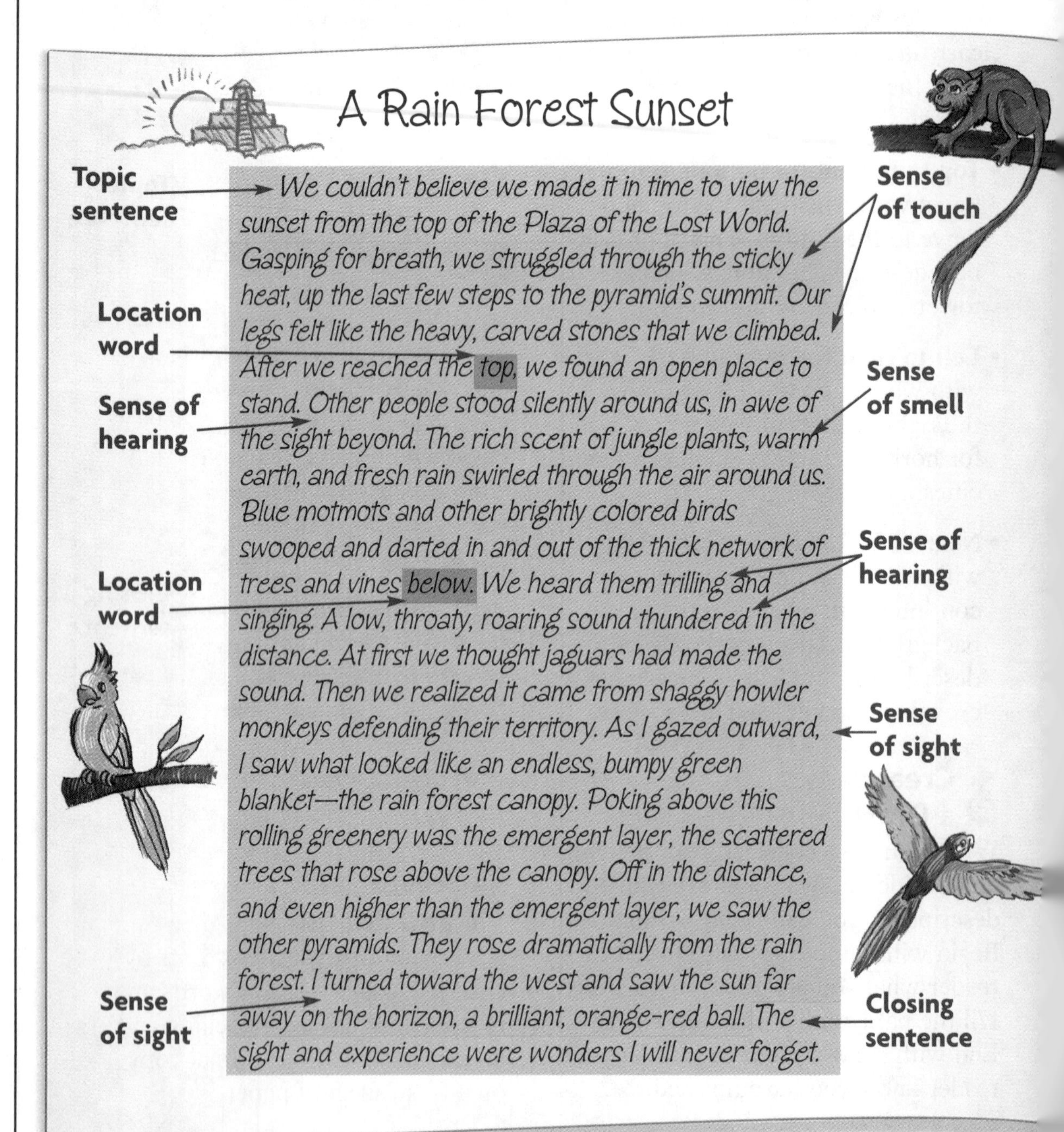

A Rain Forest Sunset

We couldn't believe we made it in time to view the sunset from the top of the Plaza of the Lost World. Gasping for breath, we struggled through the sticky heat, up the last few steps to the pyramid's summit. Our legs felt like the heavy, carved stones that we climbed. After we reached the top, we found an open place to stand. Other people stood silently around us, in awe of the sight beyond. The rich scent of jungle plants, warm earth, and fresh rain swirled through the air around us. Blue motmots and other brightly colored birds swooped and darted in and out of the thick network of trees and vines below. We heard them trilling and singing. A low, throaty, roaring sound thundered in the distance. At first we thought jaguars had made the sound. Then we realized it came from shaggy howler monkeys defending their territory. As I gazed outward, I saw what looked like an endless, bumpy green blanket—the rain forest canopy. Poking above this rolling greenery was the emergent layer, the scattered trees that rose above the canopy. Off in the distance, and even higher than the emergent layer, we saw the other pyramids. They rose dramatically from the rain forest. I turned toward the west and saw the sun far away on the horizon, a brilliant, orange-red ball. The sight and experience were wonders I will never forget.

USE THIS SKILL

Write a Description

Look at the photograph of the tropical rain forest below. Use your imagination as well as what you see in the photograph, and write a description of this rain forest.

TEST TIP	On a test you may be asked to write a descriptive paragraph. Narrow your focus to one topic and include as many details as you can. Use words that clearly tell about the topic and help the reader understand how something looks, sounds, and so on.

Skill 24

HOW TO

Write a Comparison/Contrast

What's the Difference?

Some animals that look the same at first glance are really very different. When you hear the word *elephant,* does a single picture come to mind? Maybe you already know that there are two species of elephants—the African elephant and the Asian elephant. They both belong to the elephant family and they have many similarities. However, if you compare these two elephants to each other, you will find many differences.

When you write a **comparison/contrast,** you describe how two things are alike (comparison) and different (contrast). It works best to compare things that are the same *type* of thing. For example, two types of elephants can be compared and contrasted because they are both elephants. There wouldn't be much point in comparing an African elephant to an oak tree. Elephants and oak trees are very different things.

After you figure out how two things are alike and different, you will have what you need to write a comparison/contrast.

Asian elephant

African elephant

STEPS IN Writing a Comparison/Contrast

Use these steps to write a comparison/contrast.

> **TIPS**
>
> ➤ To let your reader know about similarities, use signal words such as **also, both, like, just as, in the same way,** and **similarly.**
>
> ➤ To let your reader know about differences, use signal words such as **in contrast, different from, unlike, whereas,** and **on the other hand.**

1 Organize Information

Once you have gathered information about your subjects, you can organize it by using a graphic organizer such as a T-chart or a Venn diagram. With a T-chart, you sort through the similarities and differences in two separate columns. With a Venn diagram, you list the differences between two topics in the main regions of the two circles and the similarities in the overlapping region. The Venn diagram below puts information about two kinds of African elephants into categories.

Venn diagram

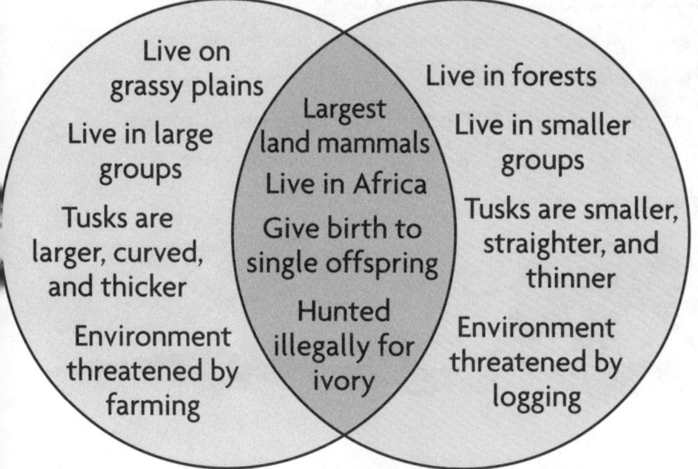

2 Decide on a Plan

Figure out how to present the differences and similarities. Use one of three approaches:

1) Describe how both topics are similar and then how they are different.

2) Discuss each topic in a separate paragraph. Then point out similarities and differences in another paragraph.

3) Discuss similarities and differences feature by feature.

3 Write Paragraphs

Introduce the two subjects at the beginning of your writing. Let the reader know you will be comparing and contrasting subjects. Begin each paragraph with a topic sentence. Help the reader follow your plan.

EXAMPLE OF **Writing a Comparison/Contrast**

Read the following passage. Pay attention to the words that signal similarities and differences. Notice how information is organized in the two paragraphs.

Asian Elephants and African Elephants

Topic sentence →

How they are similar ↗

Asian and African elephants share many features that help them survive in their environments. Both have thick, dry skin that protects them from the blistering sun. Their ears help them cool themselves in hot weather. Both kinds of elephants have an interesting feature called a trunk—a long, flexible organ that can wrap around objects such as tree branches. The trunk is also used for breathing and drinking. Both African and Asian elephants have tusks, large incisors that they use for different tasks including digging for water, scraping tree bark, and moving fallen trees. Asian and African elephants both rely on strong, bulky legs to support their tremendous weight. Both kinds of elephants usually move quietly and gracefully.

Topic sentence →

How they are different ↗

Although Asian and African elephants have similarities, they belong to two distinct species and have many physical differences. The bodies and ears of African elephants are larger in size. In fact, the African elephant is the world's largest land animal. The ears of the Asian elephant are smaller than those of the African elephant and they turn forward instead of backward. While the African elephant has two fingerlike projections on the end of its trunk for picking up things, the Asian elephant has only one. Both male and female African elephants grow tusks, but only the Asian male grows tusks. Finally, the backs of African elephants curve inward while the backs of Asian elephants curve upward.

USE THIS SKILL

Write a Comparison/Contrast

Use the information listed below to write a comparison/contrast. Before writing, create either a T-chart or a Venn diagram to organize the information.

Asian Elephants
Gray or brown skin
Head is the highest part of the body
Hind feet have four toenails
Eat grasses, bark, roots, shrubs, and tree parts
Always live near a water supply
Use trunk for eating and drinking
Females live in family groups
Females give birth to one calf at a time
Mothers care for young for years

African Elephants
Gray skin
Shoulders are the highest part of the body
Hind feet have three toenails
Eat grasses, fruit, bark, and leaves
Always live near a water supply
Use trunk for eating and drinking
Females live in family groups
Females give birth to one calf at a time
Mothers care for young for years
World's largest land animals

TEST TIP

When a test essay question asks you to compare and contrast something, take the time to organize your ideas. Think about and write ways the things are the same or different. Then use topic sentences to organize your response to the test questions.

Skill 25

HOW TO

Write about a Process

Clean Water

Think of all the places you see water every day. Water can be found as mist in the air and rain falling from the sky. It can be seen spilling over fountains and shimmering in lakes, rivers, and oceans. Whenever you turn on the faucet, you get water. Have you ever wondered where it comes from? What does it take to get clean, pure water to drink and use?

Most water used in our homes comes from lakes and rivers. Before it is purified at water treatment plants, water can contain fish, plants, debris, soil, decaying matter, and chemicals that could be harmful to people. Water can become polluted when the ground or soil is contaminated with chemicals used to kill insects, industrial waste from factories, household cleaners, and other hazardous materials. In order to make water clean enough to drink or use, water treatment plants take it through a carefully controlled process.

A **process** is a series of steps taken to make or do something. When you describe a process, you give a detailed explanation of how something happens or how something is done. When you tell how to do something, you provide the steps of the process in the correct sequence, or order.

STEPS IN Writing about a Process

Follow these steps to write about how to do something, how something is done, or how something happens.

1 Decide on a Title

Think of a short title that sums up the process you will explain in a clear way.

2 Think about Steps in the Process

Think carefully about the steps of the process and the order in which they happen. A good way to do this is to visualize, or picture in your mind, the process happening from start to finish.

3 Take Notes

As you visualize the process happening, take careful notes to briefly describe the steps. Use words that will help you explain not only what is being done but also what doing the process looks like, sounds like, and so on. You may find it helpful to arrange your notes in the same order that the steps occur.

TIP **Think about your audience before writing about a process. For whom will you be writing? Think about what your audience might already know about the process. This will help you to decide what details or directions you will need to include.**

4 Use Signal Words

Using your notes, explain the steps of the process in the correct sequence. Include signal words to make your explanation clear to the reader. These are some of the signal words that can help make the order of your steps clear.

first	before
second	during
next	after
then	later
finally	now
last	

5 Review Your Writing

Reread your process writing. Is there any step that is not needed? Did you leave any step out that you should include? Are your steps given in the correct order? Make any necessary changes. Then write a neat, final copy of your process writing on a fresh sheet of paper.

EXAMPLE OF Writing about a Process

Read below to see how one student wrote about the process of purifying water.

Title

From the River to Your Glass

Before water comes through your faucet at home, it goes through an important process to make it clean and ready to use. First water flows from rivers into treatment plants, and large objects such as plant and fish life are filtered out of it. Then water is clarified. During this stage, smaller objects or particles are removed by allowing them to settle to the bottom. Next chemicals are added to remove the smallest particles from the water. After the filtering stage, chemicals such as chlorine or ozone gas are added to disinfect the water. This removes any remaining bacteria or organisms that may cause diseases. Finally water may have fluoride added to it. Fluoride is a chemical that is sometimes put into drinking water to help you have healthy teeth. The last step of this process sends the purified water through a network of pipes out of the treatment plant and into your home.

Signal words

Steps in order

Signal words

Signal word

Signal word

Signal word

Water treatment plant

USE THIS SKILL

Write about a Process

Read and complete the following.

Look at the diagram below, which shows the process of the water cycle from the time water falls from the clouds until it is returned back to the rivers. After studying the diagram, write an explanation of this process.

Water Supply, Use, and Treatment Cycle

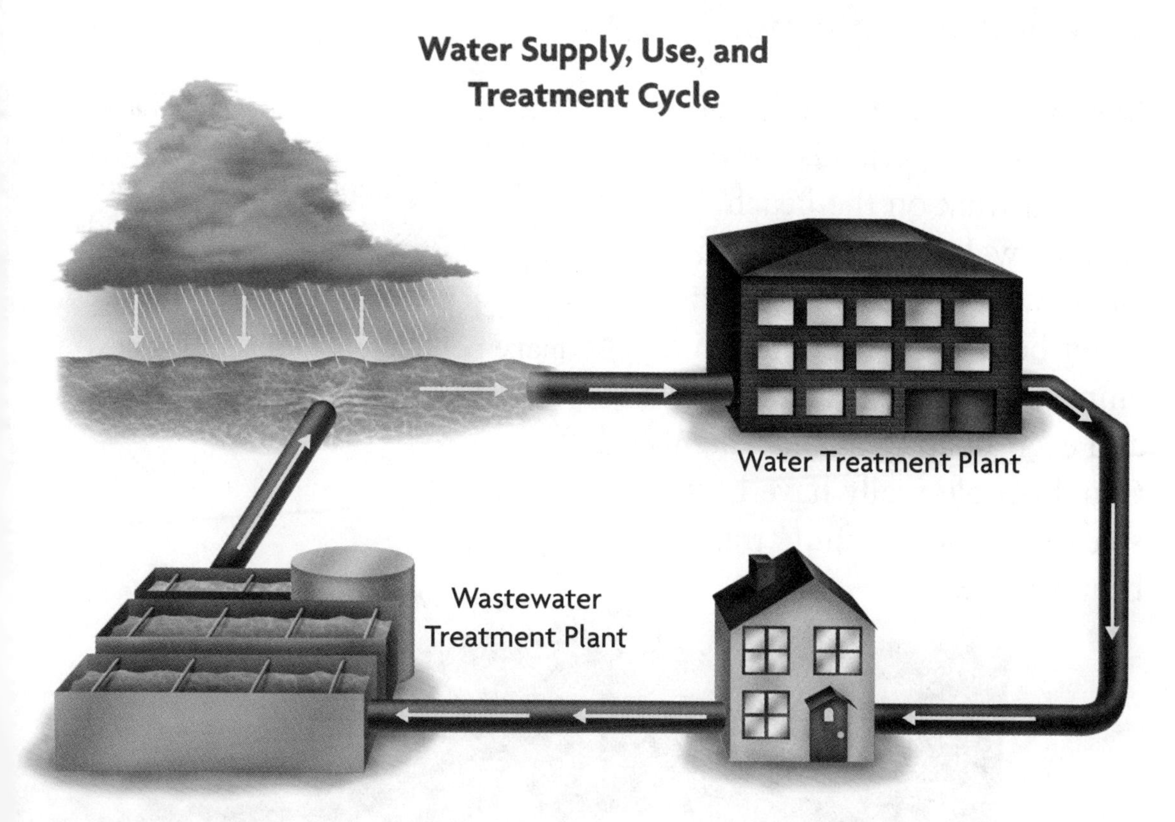

TEST TIP	When you answer test questions about a passage that explains a process, make sure you carefully note the order in which events are told. This will help keep your thinking clear about what is happening in the passage.

Skill 26

HOW TO

Write about a Problem and Solution

Solution to Pollution

One warm autumn afternoon, Miguel and his friends were taking a walk on the beach. As they walked, they noticed candy wrappers, empty plastic water bottles, plastic bags, and empty soda cans in the sand and even floating in the water. Could people really have been so careless as to pollute this beautiful beach?

The next day at school, Miguel asked his science teacher about the problem of water pollution. She told Miguel that there were many different kinds of pollution, and she suggested that Miguel do some research and then make pollution the topic of his next science essay assignment.

Water pollution

Water Pollution

Chemical Pollution

- Chemicals are dumped into the ocean.
- These chemicals can harm plant and animal life.
- Example: Factory waste containing mercury was dumped into the ocean, it got into the fish, people ate the fish, people got mercury poisoning, U.S. government banned dumping waste containing mercury.

Solid Pollution

- Tons of solid materials are dumped into the ocean.
- Plastic does not decompose.
- Sewage is often dumped by industries and people.
- Solid pollution harms wildlife.

Oil Spills

- Oil tanker ships can leak oil into the ocean.
- Oil spills harm all kinds of wildlife.

Runoff

- When rain falls, the rain runs off the land into lakes and streams.
- Sometimes runoff washes waste, fertilizers, and pesticides into the water.

Miguel began looking for sources of information about pollution. He took some notes on what he found in library books and on the Internet. He was surprised by how many different kinds of pollution were damaging the environment. Miguel soon realized that pollution was a very large, broad topic. He needed to narrow the focus of his science essay project to a specific current event, problem, or issue. Thinking about the garbage on the beach helped Miguel decide to do a problem-solution essay on pollution in the ocean.

Writing about a **problem and solution** is one type of persuasive writing. **Persuasive writing** is writing that uses opinions and evidence to try to convince people to think a certain way or take a certain action. In persuasive writing, you identify and describe a problem, and then you suggest possible solutions to the problem.

STEPS IN Writing about a Problem and Solution

Follow these steps to write about a problem and solution.

1 State the Problem

Identify the problem and think about why the problem is important. Think about who or what the problem affects.

2 Research the Causes

Do research on the causes of the problem. Use as many resources as you can find, such as reference books, encyclopedias, current magazines and newspapers, and Internet sites. Take notes on what you find, making sure to cite all resources.

3 Think of Possible Solutions

Brainstorm a list of possible solutions to the problem. Organize your ideas by making charts, outlines, and lists of your ideas. Plan to write a passage about each possible solution.

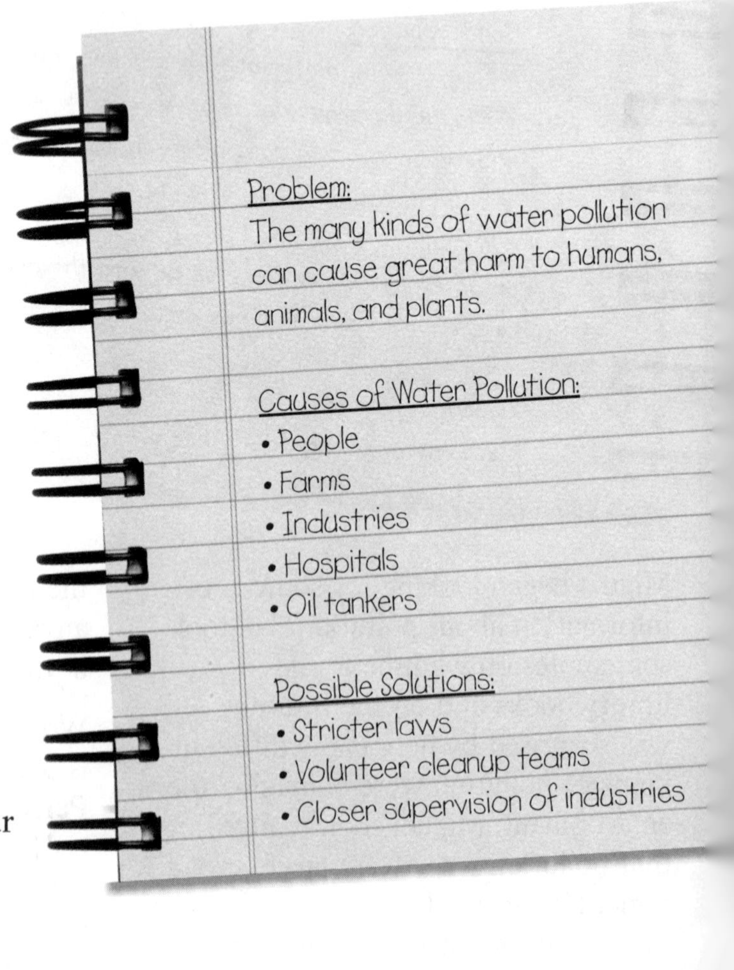

Problem:
The many kinds of water pollution can cause great harm to humans, animals, and plants.
Causes of Water Pollution:
• People
• Farms
• Industries
• Hospitals
• Oil tankers
Possible Solutions:
• Stricter laws
• Volunteer cleanup teams
• Closer supervision of industries

4 Write a Draft

Follow your notes, outline, or charts as you write a first draft of your problem-solution writing. Make sure your writing has three sections: an introduction, the suggested solutions, and a conclusion.

- Your introduction should grab your readers' attention and clearly state the problem. Give facts and examples in the introduction.
- Carefully explain your ideas for what can be done to fix the problem. Try to be as specific and detailed as possible so that your readers will know exactly what can be done.
- Your conclusion needs to contain a summary of the problem and the solutions. It should end with an appeal for action, reminding the readers what you want them to do.

Parts of a Problem-Solution Writing:

1. Introduction: Grab readers' attention; identify the problem.
2. Solutions: Write a paragraph for each solution.
3. Conclusion: Sum up the problem and the solutions. End by telling what should be done to fix the problem.

TIP **When writing to persuade someone, always give good reasons, solid facts, and clear examples to support your opinions.**

5 Review, Edit, and Publish

Reread your essay. Revise your writing by asking yourself these questions.

- Is the problem clearly presented?
- Does the introduction grab the readers' attention?
- Are the solutions clearly given?
- Does the information stick to the topic?
- Is the conclusion clearly worded? Does it clearly state what the reader should do?

Check your writing for correct spelling, capitalization, and punctuation. Once you have revised it, write a neat, final copy.

EXAMPLE OF Writing about a Problem and Solution

As you read Miguel's problem-solution essay that was printed in his school's newspaper, notice how he organized his writing.

Introduction

Can Coquina Beach Be Saved?

What's that touching your ankle as you stand in the foamy waves of our lovely Coquina Beach? Is it seaweed? No, it's the fingers of a plastic medical glove! Yes, this disgusting kind of pollution is exactly what my friends and I discovered last week as we walked along the shore.

State problem

Examples

Here are just a few examples of the pollution we found on Coquina Beach: adhesive bandages, medical syringe and needle, plastic bags, candy wrappers, empty bottles and soda cans, cigarette butts, and an empty bottle of suntan oil. All of us need to do something now if we are to have a beautiful beach to enjoy in the future. What can we as individuals do?

Solutions

One thing we can do to clean up the beach is to join a volunteer cleaning group. Businesses, churches, schools, and other organizations can sponsor cleaning groups that will take turns doing a once-a-week "beach sweep" to pick up trash. If enough groups are formed, no one group would have to do all the work.

Another solution is for the city to place more trash cans along the beach. If people do not have to walk so far, maybe they will make the effort to throw trash into a trash can instead of on the sand.

Finally, everyone needs to take care of their own trash when they visit the beach. If everyone simply cleans up his or her own little part of the beach, then no one else will have to do it for them.

Conclusion

Pollution threatens the beauty of our beach as well as the safety of the people and animals that swim in the ocean waters. We can all make a difference by joining volunteer "beach sweep" groups, by putting more trash cans on the beach, and by cleaning the areas of beach we use. Let's take back our beach and make it beautiful!

USE THIS SKILL

Write about a Problem and Solution

Think about the areas in which you live or visit. What kinds of pollution do you see? Choose one type of pollution from the list below, and use this as a topic to write about a problem and solution.

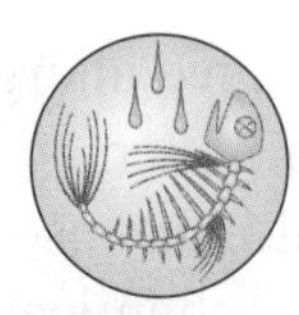

Water pollution: Solid waste and chemical pollution of lakes, streams, and oceans

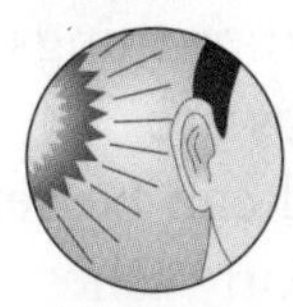

Noise pollution: Loud, constant noise made by things such as planes, cars, buses, trains, factories, and construction

Air pollution: Pollution of the air caused by vehicles, factories, and homes

Ground pollution: Pollution of the soil or ground made by litter, industrial waste, chemical disposal, and medical waste

TEST TIP	If you are asked to write about a problem and solution on a test, be sure to consider your audience before you plan your answer. Who will read what you have written? What would they care about? Focus your points on what would matter most to your readers.

Skill 27

HOW TO

Use the Internet

Food Allergies: Finding the Facts

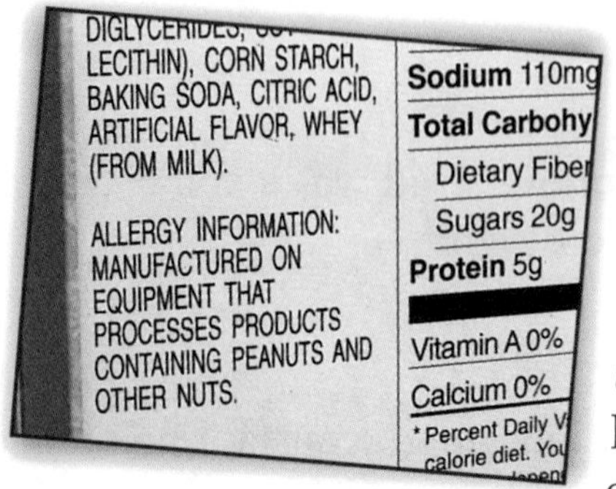

Do you know anyone who is allergic to peanuts? Are you allergic to any foods such as milk, eggs, or shellfish? Many people, children as well as adults, have allergic reactions to certain foods.

An **allergic reaction** happens when a person's immune system tries to defend the body from a substance. Things that are harmless to most people, such as certain foods or cleaning products, can cause allergic reactions in others. The immune system releases chemicals into the bloodstream that fight the substance, starting a chain reaction of responses meant to help protect the body. However, these immune responses, instead of protecting the body, can cause sickness and irritation.

People can have allergic reactions to pollen, dust, animal hair, insect stings, foods, chemicals, and many other things. Food allergies are common and can be very serious, causing illness, skin rashes, difficulty breathing, and other health problems. People who are allergic to a certain food must be very careful not to eat that food, even in small amounts. It is a good idea for people with allergies to learn as much as they can about their allergies and how to prevent serious allergic reactions.

You can use the Internet to find information about food allergies. The **Internet** is a computer network that can connect your computer to electronic resources all over the world. You can learn how to do research on the Internet by following the steps on the next three pages.

STEPS IN **Using the Internet**

1 Choose a Search Engine

Search engines are programs that help you search the Web, or World Wide Web. Ask your teacher or librarian to help you find a search engine to use. When you type the URL, or *U*niform *R*esource *L*ocator, for a search engine and press "Enter," the search engine will open and be displayed on your screen. You are now ready to look for information on the Internet.

2 Use Keywords

Decide on the best **keywords** to use for your search. These should be words related to the subject. For example, if you wanted to find food allergy information written for young people, you could use the words *Food allergies* and *Kids.* Type keywords into the search box and click on the search button.

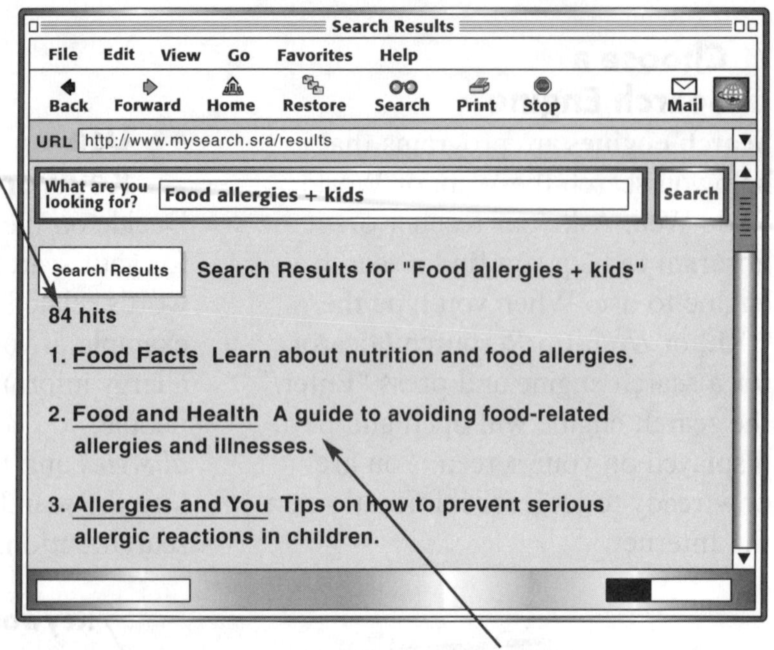

3 Select and Read Pages

The search engine will give you a list of Web pages based on your keywords. The number of Web pages found is shown on the screen as the number of "hits." There could be just one or two pages, or there could be hundreds! Look at some of the Web pages by clicking on their names. If you don't find what you want, try a new search using different keywords. Some pages will be too difficult for you to read, while others may not provide facts about the subjects you are investigating. Keep searching for pages that have information you can use.

TIP **As you compare Web sites, keep in mind that some information on the Internet may not be correct. It is a good idea to double-check facts in another source, such as an encyclopedia.**

4 Ask an Adult for Help

The Web has millions of pages. Some of these pages aren't meant for people your age. If a Web page has a warning at the top of the page or if there is a warning in a list of Web site descriptions, do not open or read the page. Never give information about yourself over the Internet unless an adult you know and trust is helping you. Always ask your teacher, librarian, or another adult for permission and help when using the Internet.

> **TIP** Look for helpful links on Web pages. Links may lead you to more information. Links are often highlighted in a different color, and may be listed along the side or bottom of a Web page.

5 Save Helpful Information

When you find a Web page that contains useful information, you may want to save it to refer to later.

- You can print entire Web pages, or select and print text, graphics, or images from helpful sites.
- You can save one section of text from a Web site by copying it and moving it to a file, then saving the file in your hard drive or on a disk.
- You may want to save the entire Web site by "bookmarking" the site or adding it to your computer's "Favorites" list. Ask your teacher or a librarian for help.

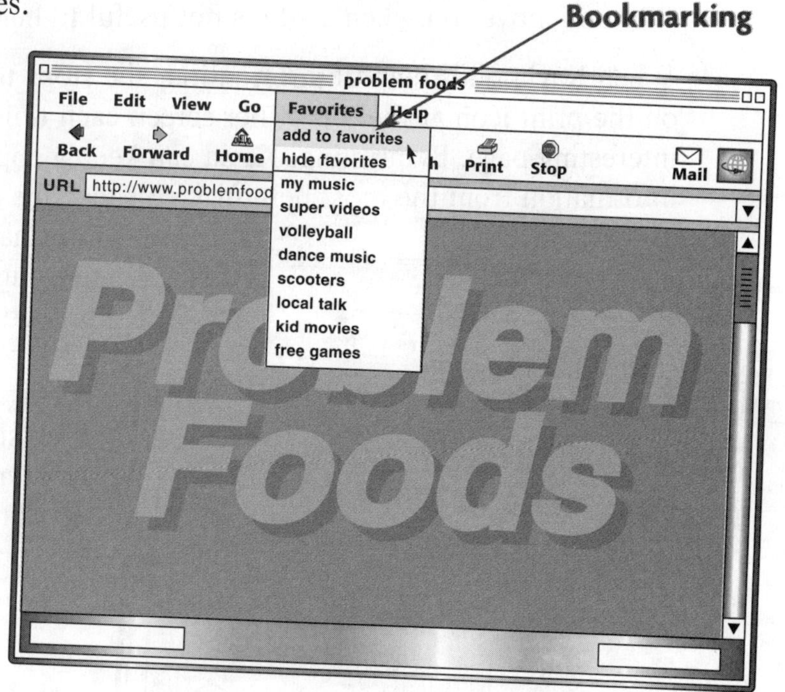

EXAMPLE OF Using the Internet

Follow one student's journey through the Internet. See how she finds information about food allergies.

- While doing research on the Internet, Kayla uses a search engine and the keywords *food allergies + kids.* Her search brings up many Web sites containing information about food allergies.
- She clicks on a few of the sites listed, but some of the sites are difficult to read. Other sites contain information that Kayla does not need. She keeps clicking on different sites and moves on when a site is not useful to her.

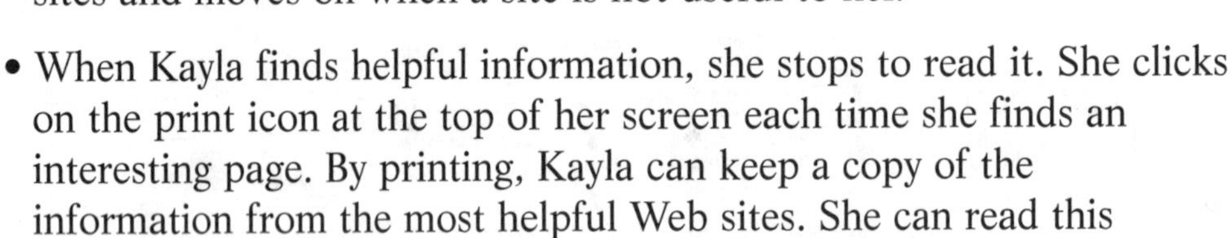

- When Kayla finds helpful information, she stops to read it. She clicks on the print icon at the top of her screen each time she finds an interesting page. By printing, Kayla can keep a copy of the information from the most helpful Web sites. She can read this information more thoroughly later. Each printed copy has the site's URL on the page. If Kayla needs to go back to the Web site another time, she can type in the URL to return directly to the site.

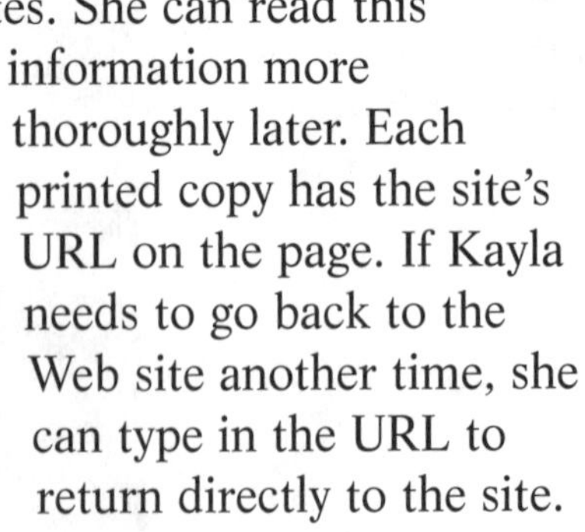

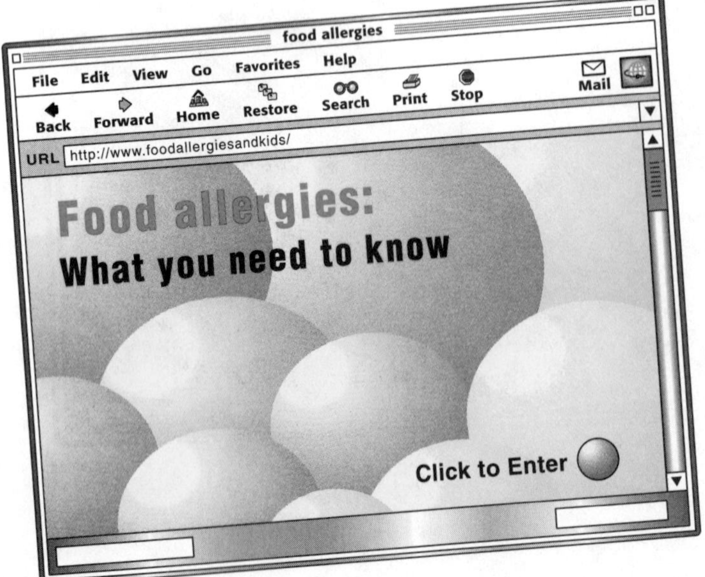

USE THIS SKILL

Use the Internet

Use the Internet to find information about one of the topics listed below. Ask your teacher if you need help finding a search engine to use.

- Many young people have serious allergies to peanuts. Learn the symptoms of peanut allergies and what can be done to treat these symptoms.
- Citrus fruits cause allergic reactions in some people. Find out what fruits are classified as citrus fruits and what products people with citrus allergies must avoid.
- Find information about ways that doctors and allergists can treat some food allergy symptoms, such as hives and asthma.
- Look up the word *anaphylaxis.* Try to find several paragraphs of information about anaphylaxis. Write the definition of the word, and take some notes from the paragraphs you find. Give one or two examples of anaphylactic reactions.
- Find out the difference between being allergic to a food and being "intolerant" of a food. Write down some of the key points of the information you find.
- Learn more about the immune system by searching allergy Web sites for details about immune response. Write a list of some of the important functions of the immune system.

TEST TIP

Some tests may ask you to identify keywords in a reading passage. Before answering questions about keywords, carefully read the passage on the test. Keywords will be important words related to the subject.

Skill 28

HOW TO

Use Reference Sources

Building the Brooklyn Bridge

Completed in 1883, the Brooklyn Bridge was a major achievement in science and architecture. In addition to being the longest suspension bridge of its time, it was the first suspension bridge to use steel cable wires that were suspended from tall towers. When it was finished, the Brooklyn Bridge's stone towers were the tallest structures in the United States.

Engineers who design bridges need to have an understanding of the forces at work on the cables, the towers, and the bridge's deck. They understand the science of **statics,** a type of science that deals with maintaining equilibrium. **Equilibrium** is the state in which forces acting on an object are exactly balanced.

In a **suspension bridge,** the roadway is suspended from enormous steel cables.

The Brooklyn Bridge

In the Brooklyn Bridge, each cable is about the diameter of a telephone pole and is made up of more than 5,000 steel wires. The roadway hangs 135 feet above the water, allowing ships to pass below it. The four cables pass through two towers and are anchored in solid concrete blocks at either end. These enormous concrete blocks were designed to have enough mass to offset the pull of the cables. The two towers have enough mass to support the huge downward pressure of the cables, the bridge deck, and the vehicles that travel across the bridge. The design of the Brooklyn Bridge is an amazing example of equilibrium at work.

The Brooklyn Bridge's engineer was a German immigrant named John Augustus Roebling. When he died in 1869, his son Washington Roebling took over the project. After fourteen years of difficult labor and many construction problems, the bridge opened in 1883.

How exactly was the bridge built? To find out more about the Brooklyn Bridge, you can use reference sources. **Reference sources** are materials that give information about different subjects. Some of the many different kinds of reference sources are:

Newspapers and Magazines

- Include up-to-date information about current events
- Published daily, weekly, monthly, or quarterly

Books, Almanacs, Encyclopedias, and Atlases

These are found in libraries, bookstores, and even online. There are books that provide detailed information about almost any topic you can imagine.

Almanacs are published yearly and give up-to-date statistics and other information on a variety of different subjects.

Encyclopedia articles give general information about places, things, people, and events. Look for online encyclopedias, or check your library for different printed volumes and encyclopedias on CD-ROM.

Atlases are books of maps. Use an atlas to find out more about places in the world.

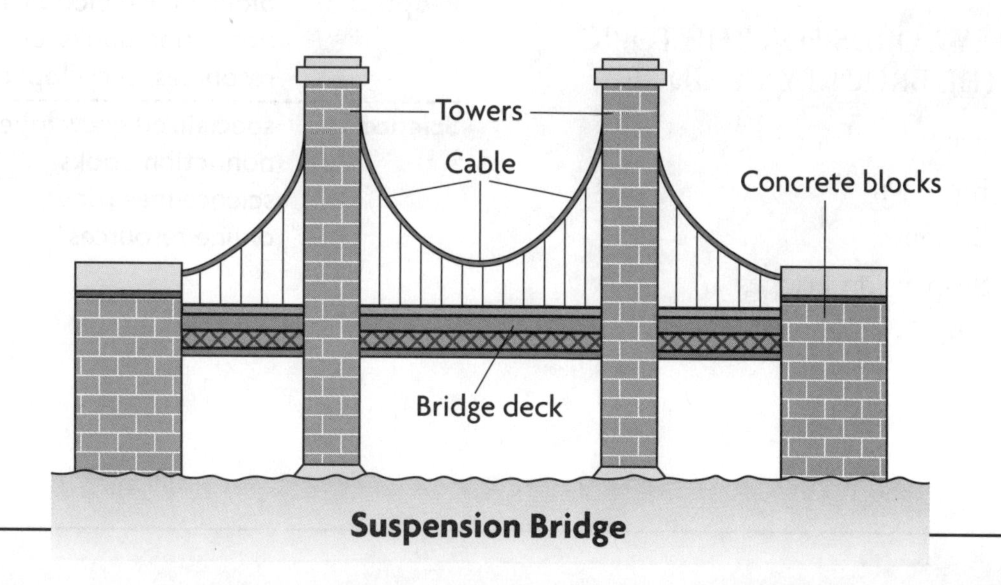

Suspension Bridge

STEPS IN Using Reference Sources

Follow these steps to learn how to use reference sources.

1 Choose a Topic

What kinds of information are you looking for? Write some questions you have about a topic.

2 Identify Keywords

Keywords are main words or phrases that have to do with your topic. Use keywords to help you focus your search and save time. Think of some phrases, names, or words that relate to your topic. You can look up these keywords in the different reference sources you choose. Online search engines will ask you to type in a keyword to help you find appropriate Web sites.

KEYWORDS FOR THE TOPIC "THE BROOKLYN BRIDGE"

- Brooklyn Bridge
- Brooklyn
- Bridges
- Suspension bridges
- John Roebling

3 Decide What Kinds of Sources to Use

Think about which kinds of reference sources will give you the information you need. Because Web sites provide huge amounts of information on just about any subject, the Internet is a good resource to use. Here are a few other resource suggestions:

TOPIC	RESOURCES
Current events	newspapers, magazines, online resources
Historical events	nonfiction books, encyclopedias, online resources
Places	atlases, almanacs, online resources
People	biographical dictionary, nonfiction books, online resources, encyclopedias
Science	specialized encyclopedias, nonfiction books, science magazines, online resources

4 Search and Take Notes

Take notes, including the title and author or Web address, for each source you use. As you find answers to your questions about your topic, write them in your notebook along with the source and page number where you found them. To help you find your way through all the information a source offers, use the tools provided with each source:

- **Table of contents.** Many sources have a table of contents at the beginning. In books, the table of contents lists chapter titles and the page numbers on which they begin. Magazines give the page numbers of articles. Do a quick check of a source's table of contents to decide if the source will be useful.
- **Index.** Many reference sources have indexes on their last few pages. An **index** is a list of topics that are discussed in a source. The topics are listed alphabetically by keywords or a person's last name. Nonfiction books and encyclopedias also have indexes. Skimming an index is a good way to figure out if a source will meet your needs.

Bartholdi, Frederic Auguste **B: 120**

Hunt, Richard Morris **H: 450**

Liberty Island **L: 233**

New York City (A visitor's guide) **N: 338–339** *with picture*

Statue of Liberty S: 872 *with pictures and map*

"120" is the page number

"S" is the volume letter

Sample encyclopedia index

- **Glossary.** Provided at the back of many nonfiction books, a **glossary** gives definitions of important words used in the book. Like a dictionary, it lists words alphabetically.
- **Copyright/title page.** The copyright page in the very front of a source gives publication information about a book or other reference source. Look at the copyright page to find out when and where a source was published.
- **Guide words.** Sources such as encyclopedias use guide words. These appear at the top of each page and tell you what articles appear on that page. Use guide words to quickly find a topic as you skim through a reference source.

EXAMPLE OF **Using Reference Sources**

Rasheed is preparing to write a report about the Brooklyn Bridge. Read below to see how he uses reference sources in his research.

The Building of the Brooklyn Bridge–
History and How It Works
I'll need to find out about the history of the bridge and the technical planning involved. These keywords should help me find both.
Brooklyn Bridge
Suspension Bridges
~~Brooklyn~~ (too general)
John Roebling
This is a topic about both a historical event and the technology of building a suspension bridge. I think there are a lot of different sources I could use. Possible sources:
Web sites
Encyclopedia
Nonfiction books about Brooklyn, bridges, and the Brooklyn Bridge
Magazine articles?
First I'll try the encyclopedia to get a general idea about my topic. Then I'll use what I found there to search the Internet. Finally I'll see what kinds of nonfiction books the library has.
- ***Internet sites***

The Brooklyn Bridge–Gateway to a Century
Brooklyn Bridge Historical Society
- ***Encyclopedia***

Short article about Brooklyn Bridge
Longer, more detailed article about bridges in general
- ***Computer card catalog***

Hundreds of books on bridges; many about the Brooklyn Bridge
Internet: *http://www.pbs.org/wgbh/buildingbig/bridge/susp_forces.html–*
great site because it gives detailed history, as well as statistics and explanation of the forces at work on a suspension bridge

USE THIS SKILL

Use Reference Sources

Look at the two topics below. For each one, list the types of reference sources you think will help you find information. Then use reference sources to answer the questions about each topic.

Modern Structures
Bridges, tunnels, and skyscrapers

1. What materials were used to build the Golden Gate Bridge in San Francisco?
2. Was the science of statics used in the design and building of the Lincoln Tunnel in New York City?
3. What types of materials are used to build skyscrapers such as the Sears Tower in Chicago?
4. Other than the triangle, what other shapes do engineers use when building structures for height and strength?

Historical Structure
The Pyramids at Giza, Egypt

5. What materials were used in building these pyramids?
6. How were these pyramids designed and built?
7. What shapes were used in building these pyramids?
8. Have the techniques used to build these pyramids ever been used to build modern structures?

TEST TIP

Some tests, called open-book tests, allow you to use books or other sources during the test. Use the table of contents or the index of each book to help you find the sections you need. Remember to work quickly and to look up only the topics that you need to answer the test questions.

Skill 29

HOW TO Write a Report

Animal Communication

Grace wondered what she could write about for her science report. She had to write something about animal behavior, but animal behavior seemed like such a big topic. Grace started by thinking about what she already knew about animal behavior and what she wanted to know.

Her dog Leo did lots of interesting things. To invite another dog to play, he would stretch his legs in front of him and bend his body to the ground. If he wanted to play with her, he would nudge her with his favorite plastic bone. When he wanted to eat, he pawed at his food bowl. It really seemed like Leo communicated not just with other dogs but also with people. That was it! Grace decided she would write about how animals communicate.

One way to share what you learn about a subject is to write a report. **Reports** provide information and explanations about real events, facts, and ideas. A **research report** involves using reference sources to gather information that will be included in a report. You and others should be able to check this information by referring to articles, books, online information, newspapers, encyclopedias, and other reference sources. Reports may include diagrams, photographs, or illustrations with captions to help readers understand information.

When Grace began doing research for her report, she found out that different animals communicate in different ways. For example, she learned that wolves howl to claim territory and also to let pack members know where they are while hunting. Gorillas stick out their tongues to show anger. Chimpanzees grin widely and show their teeth when they are afraid. Birds sing to attract mates and also to mark territory. Grace thought that bee communication was the most interesting of all. A female honeybee does a "dance" that tells other worker bees where to find a patch of flowers for food.

Grace saw this illustration in an article about honeybee communication.

After a honeybee scout finds flowers that are blooming, she returns to the hive and does one of two main dances to tell other worker bees about the food.

A. *This simpler dance called the "round dance" tells other worker bees that food is very close to the hive.*

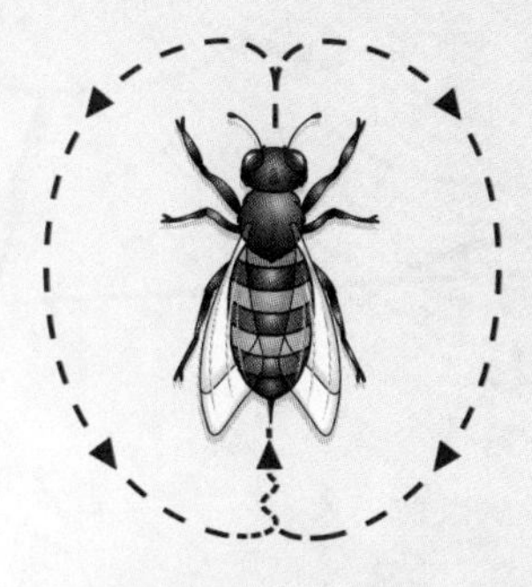

B. *This more complicated dance, called the "waggle dance," tells other bees in which direction and also how far to fly for food that is farther away. The straight line points in the direction of the food. The number of "waggles" tells the distance.*

STEPS IN Writing a Report

Use the following steps to help you write a report.

1 Choose and Narrow Your Topic

Sometimes your teacher will assign a topic. Other times you will select your own topic.

Graphic organizers can help you sort through information when deciding upon a topic for your report. Notice the broad topics and the smaller topics in the concept web below. Because animal communication is a big topic, it could be narrowed to one smaller topic such as how the prairie dog communicates. Another topic could be how different animals use just one method, such as scent, for communication.

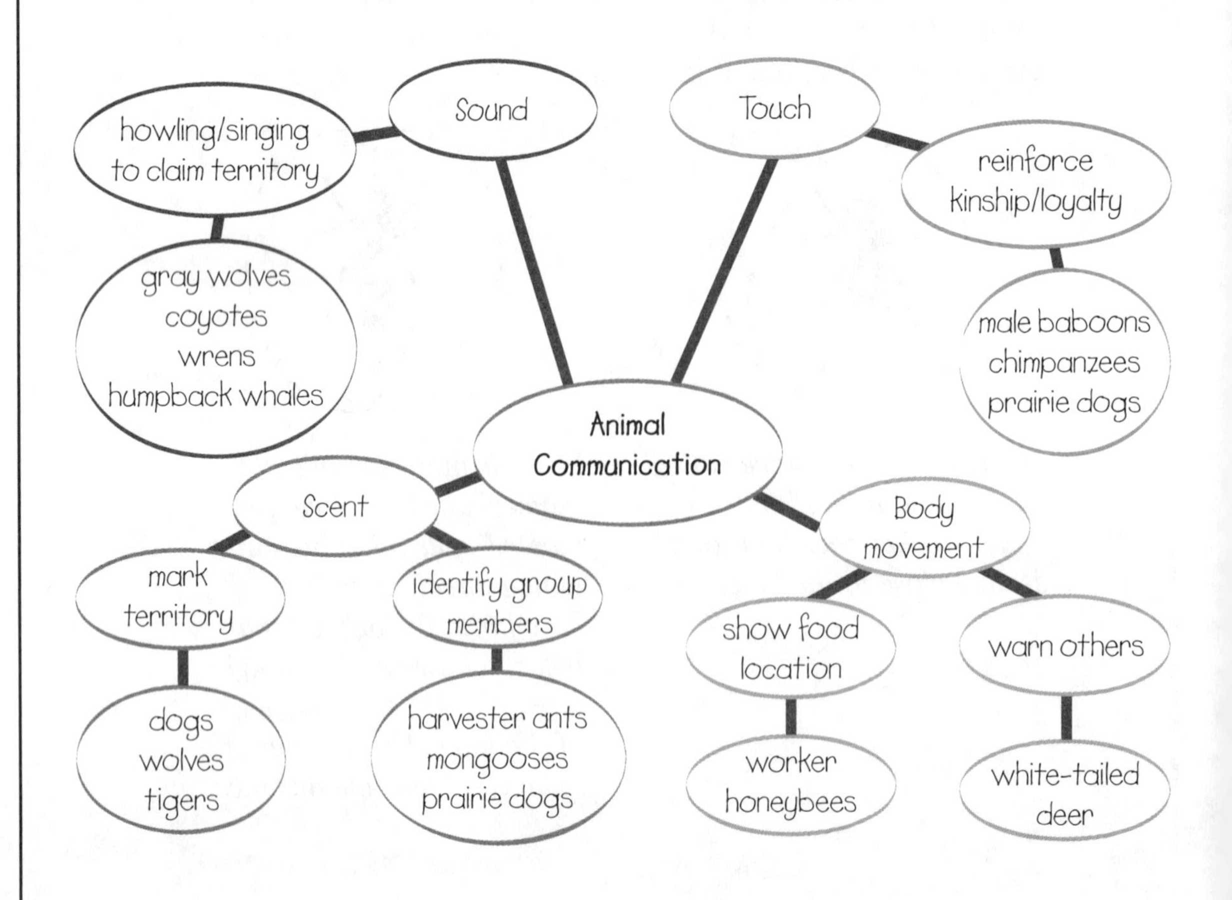

2 Gather Information

Begin collecting information about your topic by searching through books, Web sites, magazine articles, encyclopedias, and other resources. Keep track of what you learn by taking notes. Use more than one source to check facts. Keep track of each source by making a bibliography list.

How do male wolves use body movements to communicate?
- *Crouch low toward ground*
- *Lower head and tail*
- *Flatten ears*
- *Lick leader's muzzle*
- *Whine quietly*

Information note card

3 Organize Your Information

Think about the purpose of your report. Who will read your report, and what do you want them to understand? Which facts and ideas will interest them most? Organize your ideas so they are easy to follow. Try using a graphic organizer or an outline to plan the key points and the supporting information that you will use.

4 Draft Your Report

Start each paragraph with a topic sentence, then add details. As you write your first draft, don't worry too much about grammar and spelling. Include an introduction that captures your reader's interest and provides some background. Write a conclusion that summarizes your main points or invites readers to think about what they learned in your report.

5 Revise and Edit

Reread your report to see if you have provided enough facts and details to support your key points. Be sure you have given credit to any quoted sources. Correct any grammar, spelling, and punctuation mistakes. Neatly type or write a final copy, including all the corrections. Add any diagrams or illustrations that will help your reader understand the topic.

TIP Writing a report may seem like a big job, but it doesn't have to be. Start working on your report well before the due date, and break the work up into smaller tasks, such as working on one paragraph at a time.

EXAMPLE OF **Writing a Report**

Read this example of the first page of a student's research report.

Title → *Honeybee Communication* *Grace Nakamura*

Science, Grade 5

Introduction

Main topic

Have you ever seen one bee moving from flower to flower, and a little while later many more bees are buzzing around the same flowers? Honeybees have a way of communicating to one another where to find food. First they must learn the location of food by using their senses of sight and smell. Then scout bees communicate the location of food to other bees by doing a special dance.

Topic sentence

Supporting information

Worker bees, which are always female, are the food collectors for the hive. These worker bees send out bees called scouts to find food. After one finds a rich food source and collects pollen and nectar, it returns to the other worker bees. The scent of the pollen or nectar on its body tells the other bees what the flowers smell like.

Topic sentence

Supporting information

Worker bees do one of a few different dances to communicate where to find the flowers. A dance with lots of turns and frequent wiggling tells other bees that the food source is very near. A dance that has a straight line in the middle tells the other bees that the food is farther away. The direction of the straight line points the bees in the direction they must fly to find the flowers.

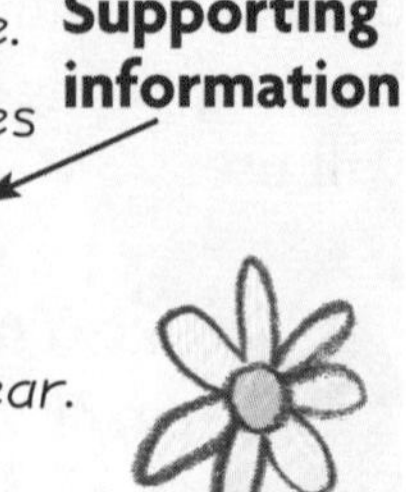

USE THIS SKILL

Write a Report

Choose one of the animals listed below, or select another animal of your choice. Write a report about how this animal communicates.

ANIMAL	TOPIC
Robins and other birds	Find out how the color of the robin's red breast sends a message to others of its species. Explore how other types of birds use colored feathers to communicate.
Bottlenose dolphins	Explain how they are able to communicate with other dolphins and also with humans.
Gray wolves	Explore wolf communication and behavior. Describe how some scientists use wolf pack interactions to explain behaviors between pet dogs and human owners.
Chimpanzees	Learn how some humans have taught chimpanzees to use words and other human symbols to communicate. Investigate the problems with making conclusions about these experiments.
Stickleback fish	Discover and describe how stickleback fish use a sequence of signals and movements to communicate with other sticklebacks.

TEST TIP

Some test essay questions require long answers. Before writing your answer, organize your ideas as you would when writing a report. As you write, begin each paragraph with a topic sentence and support it with details.

Skill 30

HOW TO

Prepare an Oral Report

Dr. Mae Jemison

When Mae Jemison blasted into orbit on the space shuttle *Endeavour* in September of 1992, she became the first African American woman in space. Being chosen as the first woman of color to be an astronaut by NASA is only one in a long list of her outstanding accomplishments. She has been a chemical engineer, a medical doctor, a teacher, a research scientist, *and* an astronaut.

Dr. Mae Jemison

Dr. Jemison has also been an inspiration to young people everywhere and says, "Don't let anyone rob you of your imagination, your creativity, or your curiosity. It's your place in the world, it's your life. Go on and do all you can with it, and make it the life you want to live."

Such a fascinating person makes a good subject for an oral report. In an **oral report,** you tell about your subject by speaking. Instead of writing a long, detailed report that others will read, you give information about your subject in person by talking to an audience. You can learn how to prepare and present an oral report by following the steps on the next three pages.

STEPS IN Preparing an Oral Report

1 Gather Information

Do research about your subject using several different sources. For example, look in an encyclopedia to find basic information such as names and dates. Other nonfiction library books often will provide more detailed information about your topic. The Internet is also a good source of information, especially if you are looking for the most up-to-date facts. Newspapers and magazines are other places to look for current information.

Take notes.
As you read through your sources, take notes on the most important facts you find. Include the source's title, author, publisher, date of publication, and the page numbers you use. If you found the information on the Internet, include the Web site address and the date you found it.

2 Organize Information

Think about how you will organize your report. Review your notes to figure out what categories you will include, such as *childhood* or *education.* Then arrange the categories in an order that makes sense. If your report is about a person's life, you might decide to organize your notes starting from the beginning of the person's life to the present day. Look at this example:

Subject of report: Dr. Mae Jemison

1) Childhood and education
2) Work with NASA
3) Other accomplishments and awards

3 Prepare Note Cards

Write key points from your notes on note cards. At the top of each card, write the main topic. Then write keywords and phrases about that topic below. Put the cards in the order you wish to present each topic. Number each card. You can use the note cards as a guide while you speak, which will help you remember what to say.

Introduction

Write an introduction on your first note card. It should tell the subject of your report. You might include an interesting quote or a question that grabs your audience's attention.

Ending

On your last card, write a very short summary of your report. Leave the audience with something interesting to remember about the subject of your report.

4 Prepare Visual Aids

Visual aids are things you can show to help illustrate your ideas and keep your audience interested. Here are some ideas for visual aids you can use:

- **A time line.** This may cover a person's life, events in history, and so on. Make it large enough so that your audience can see the labels and dates.
- **A photo, drawing, or model.** Make copies of pictures of the person or thing your report is about, or draw pictures to show what you want your audience to see. Create a model or diorama to help illustrate the subject of your report.

Decide how you will hold or display visual aids. Make sure they are large enough for the whole class to see.

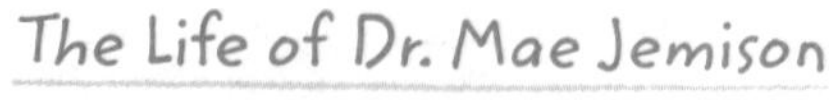

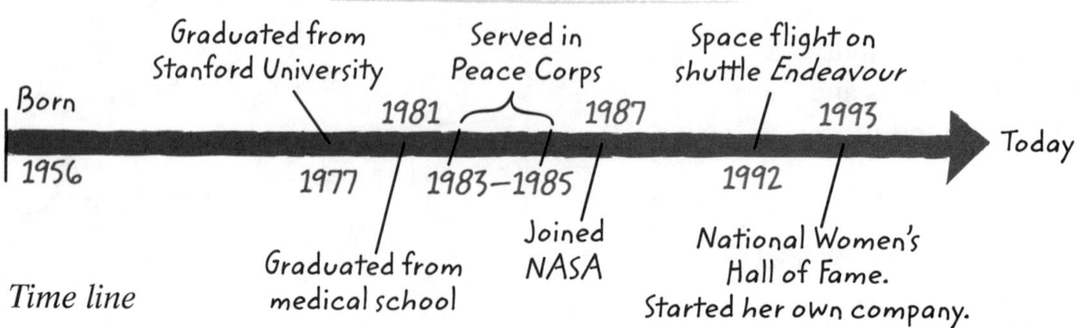

Time line

INTRODUCTION ①

Title of report: "Dr. Mae Jemison–My Favorite Scientist"
- Knew she wanted to be a scientist at age five
- First African American woman in space
- Only one of many outstanding achievements in her remarkable career

5 Practice Your Report

Give yourself a few days to practice before you present your report to the class. Practice in front of family members or friends, or in front of a mirror. Practice saying difficult names or other words. The more you practice, the more comfortable you will be in front of an audience. Ask your practice audience for ideas for improving your presentation.

As you practice, keep these speaking skills in mind:

- Start by introducing yourself and telling what or who your report is about.
- Look right at the faces in your audience. Make eye contact.
- Speak clearly and loudly enough for everyone to hear.
- Don't worry if you make a mistake. Just smile and keep going!

6 Present Your Report

Make your oral report lively by speaking directly to your audience and making eye contact with them. Your words will sound more natural and interesting to your audience if you face them as you speak. Glance quickly at your note cards, but try not to read word for word.

TIPS

- **Use a highlighter pen to mark important points you have written on your note cards. This way, you won't overlook a key idea.**
- **Use arrows or other symbols in your note cards to remind yourself when to show the audience a visual aid.**

EXAMPLE OF Preparing an Oral Report

See how one student used note cards to prepare an oral report on scientist and astronaut Mae Jemison.

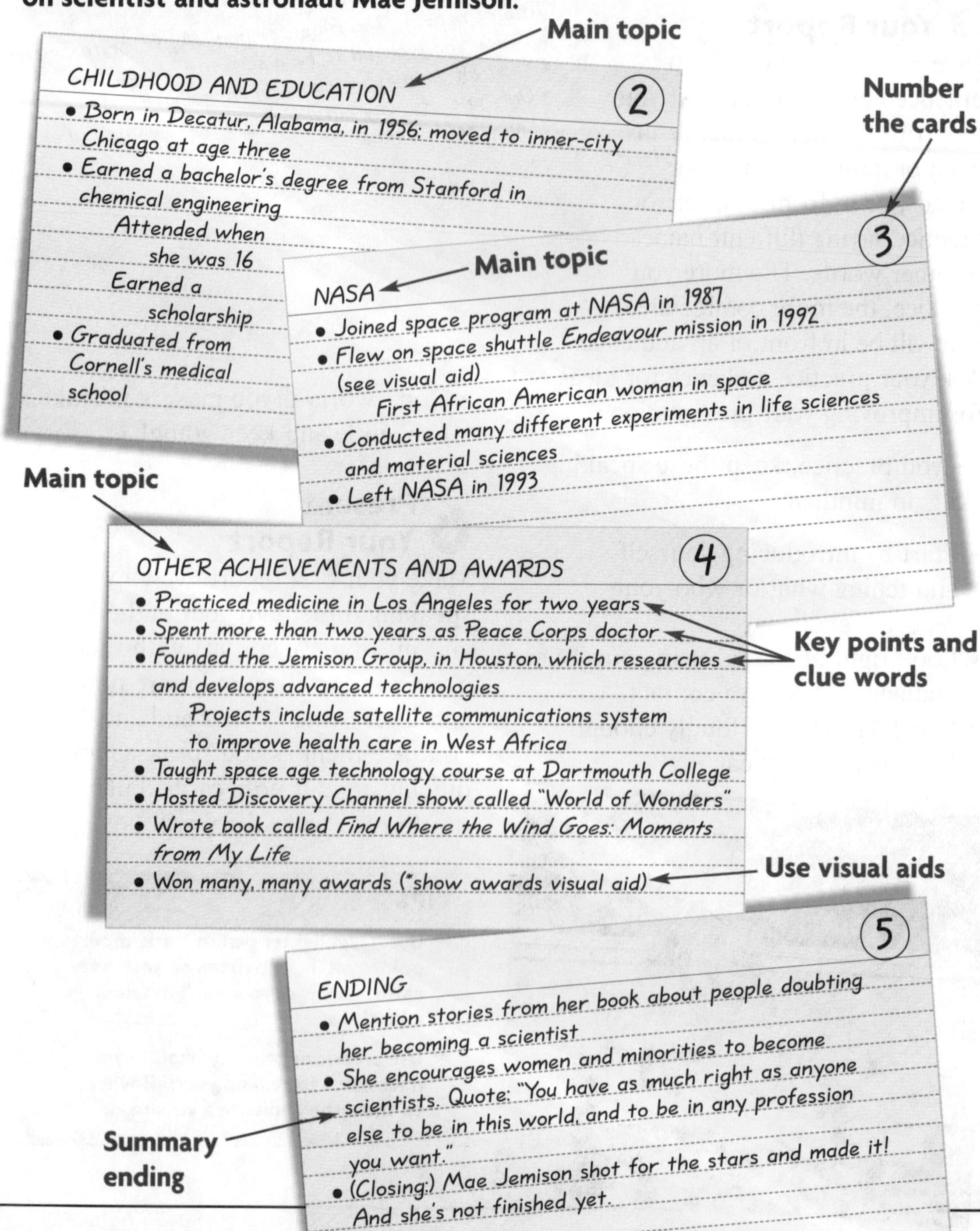

Main topic
CHILDHOOD AND EDUCATION
2
• Born in Decatur, Alabama, in 1956; moved to inner-city Chicago at age three
• Earned a bachelor's degree from Stanford in chemical engineering
Attended when she was 16
Earned a scholarship
• Graduated from Cornell's medical school
Number the cards
3
Main topic
NASA
• Joined space program at NASA in 1987
• Flew on space shuttle Endeavour mission in 1992 (see visual aid)
First African American woman in space
• Conducted many different experiments in life sciences and material sciences
• Left NASA in 1993
Main topic
OTHER ACHIEVEMENTS AND AWARDS
4
• Practiced medicine in Los Angeles for two years
• Spent more than two years as Peace Corps doctor
• Founded the Jemison Group, in Houston, which researches and develops advanced technologies
Projects include satellite communications system to improve health care in West Africa
• Taught space age technology course at Dartmouth College
• Hosted Discovery Channel show called "World of Wonders"
• Wrote book called Find Where the Wind Goes: Moments from My Life
• Won many, many awards (*show awards visual aid)
Key points and clue words
Use visual aids
5
ENDING
• Mention stories from her book about people doubting her becoming a scientist
• She encourages women and minorities to become scientists. Quote: "You have as much right as anyone else to be in this world, and to be in any profession you want."
• (Closing:) Mae Jemison shot for the stars and made it! And she's not finished yet.
Summary ending

USE THIS SKILL

Prepare an Oral Report

Choose a subject from the list below. Prepare note cards for an oral report on that person's life and accomplishments.

Women in NASA	
Sally Ride	First American woman in space
Kathy Sullivan	First American woman to walk in space
Aprille Ericsson-Jackson	First African American woman to receive a Ph.D. in Engineering at NASA
Shannon Lucid	First American woman to spend extended time on Russia's Mir Space Station
Laura Hoppe	First female INCO flight controller at Johnson Space Center
Eileen Collins	First woman to be selected as a space shuttle pilot and first woman to command a space shuttle

TEST TIP

When studying for a test, you must remember a lot of different pieces of information. One way to do this is to write key ideas, and then put your notes away and recite the ideas aloud. Do this until you can recite them without looking back at your notes.

Skill 31

HOW TO

Do a Survey

Protecting Your Body

When the weather is nice, you probably want to be outside having fun. You buckle on your skates or hop on your bike and off you go. But wait! Have you forgotten something?

Skating and biking are great ways to exercise and have fun, but they can also result in injuries. To keep yourself safe, you should wear special protective gear. A sturdy helmet can protect your head if you fall on the ground or pavement. Padding can prevent scrapes, sprains, and broken bones. In many places, there are even laws that require young people to wear protective gear such as bicycle helmets.

Protective gear is also required when playing many sports. In softball, a batter wears a special helmet with a face guard. Football players wear padding to protect their shoulders, necks, backs, and chests. Mouth guards keep their teeth and jaws safe.

Athletic shoes that offer good support help prevent ankle and foot injuries.

Do your classmates wear protective gear when they play sports or exercise? Has it helped them avoid injuries? One way to find out is to do a survey. A **survey** is a set of questions asked of a group of people. The survey can be spoken or written on paper. The steps on the next page explain how to do a survey, count the responses, and show the results.

STEPS IN **Doing a Survey**

1 Decide What You Want to Know

Write down the questions you have. For example, ***Do soccer players who wear shin guards have fewer leg injuries?*** Give your survey a title that tells the topic of the survey.

2 Decide Who to Survey

Select your survey group according to what you want to learn. If your main question is specific like the one above, you should survey only certain people, such as soccer players. If your main question is general, for example, ***What is your favorite sport?***, survey a larger group of people.

3 Write and Ask Survey Questions

- You can ask questions with specific answers, such as yes or no: ***Have you ever had a leg injury while playing soccer?***
- You can ask questions with many possible answers: ***What protective gear do you wear when you play soccer?***
- Avoid asking questions in a way that could influence the response. For example, the question ***You wear shin guards when you play soccer, don't you?*** suggests a *yes* answer.

4 Tally the Responses

Create a box next to each question where you can tally, or record, the answers.

The easiest questions to tally are those with yes/no answers. Place a slash in either the *yes* box or the *no* box to represent either a yes or no answer. If you ask questions with many possible answers, leave space or empty boxes on your response sheet for other answers.

5 Show the Results

You can show the results of your survey in several ways. One way is to determine the percentage of each response and show it on a pie chart. Another way is to graph the results. Or simply count the total responses to each question and record these numbers.

TIP **When taking a tally, organize your slashes into groups of five so that you can count them more easily.**

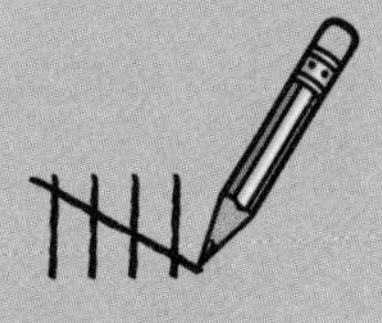

EXAMPLE OF Doing a Survey

Read this survey about sports gear and injuries.

Who you surveyed

Title

What you want to know

Responses

Sports, Sports Gear, and Injuries	Answers	Totals
25 students responded to the survey. Students are allowed to say yes to more than one answer in each section.		Out of 25
1. What kinds of sports activities do you enjoy?		
Running	𝍸 𝍸 𝍸 |	16
Batting/hitting a ball	𝍸 ||||	9
Kicking a ball	𝍸 𝍸 𝍸	15
Swimming	𝍸 𝍸 𝍸 𝍸	20
Skating	𝍸 𝍸 |||	13
Biking	𝍸 𝍸 𝍸 𝍸 ||||	24
2. What kinds of protective gear do you use when doing these activities?		
Helmet	𝍸 𝍸 ||||	14
Mouth guard	𝍸 ||	7
Shin guards	𝍸 𝍸 𝍸	15
Elbow pads	𝍸 𝍸 𝍸 ||	17
Knee pads	𝍸 𝍸 𝍸	15
Goggles	|||	3
Athletic shoes	𝍸 𝍸 𝍸 𝍸 𝍸	25
Face mask	𝍸 |	6
3. What injuries have you had while doing these activities?		
None	𝍸	5
Scrapes	𝍸 𝍸 𝍸 |||	18
Broken bones	||	2
Sprained muscles	||||	4
Loose teeth	|	1
Head injury	||	2
4. Were you wearing protective gear at the time	Yes 𝍸 |	6
of the injury?	no 𝍸 𝍸 ||	12

Results

Results: Of the 25 students surveyed, 20 said they have had injuries while doing sports activities. Of these 20 students, 6 said they were wearing protective gear at the time of the injury, and 12 said they were not.

USE THIS SKILL

Do a Survey

Read the passage below. Then write questions to survey other students. Tally their responses and show the results.

You have asked for a pair of in-line skates for your birthday. You decide to survey other students who have in-line skates to see what protective gear they use and what injuries they have had. Then you will know what kinds of gear you will need to wear when using your new skates.

Survey other students who have in-line skates. Ask them questions that will give you the information you need. Then tally their responses and show the results.

TEST TIP

You may be asked to make a survey as part of a test. Carefully read the purpose of the survey. Ask only questions that are directly related to the purpose of the survey.

Chart and Graph Skills

Skill 32

HOW TO

Make a Time Line

Geological Time Scale

Think back over your life. What major events have happened to you? When did these events occur? If you wanted to tell when important events of your life happened, you could show a visual record of them by making a time line.

A **time line** is a type of diagram used to list events that happened over a period of time. That period could be any length of time, as short as one day or longer than millions of years. Time lines are divided to represent different time spans. Labels that describe particular events are placed at the correct time period or date on the time line. A time line is a clear and easy way to chart events. It is much easier to understand the sequence, or order, of many events on a time line than it would be to read paragraphs that tell about these events.

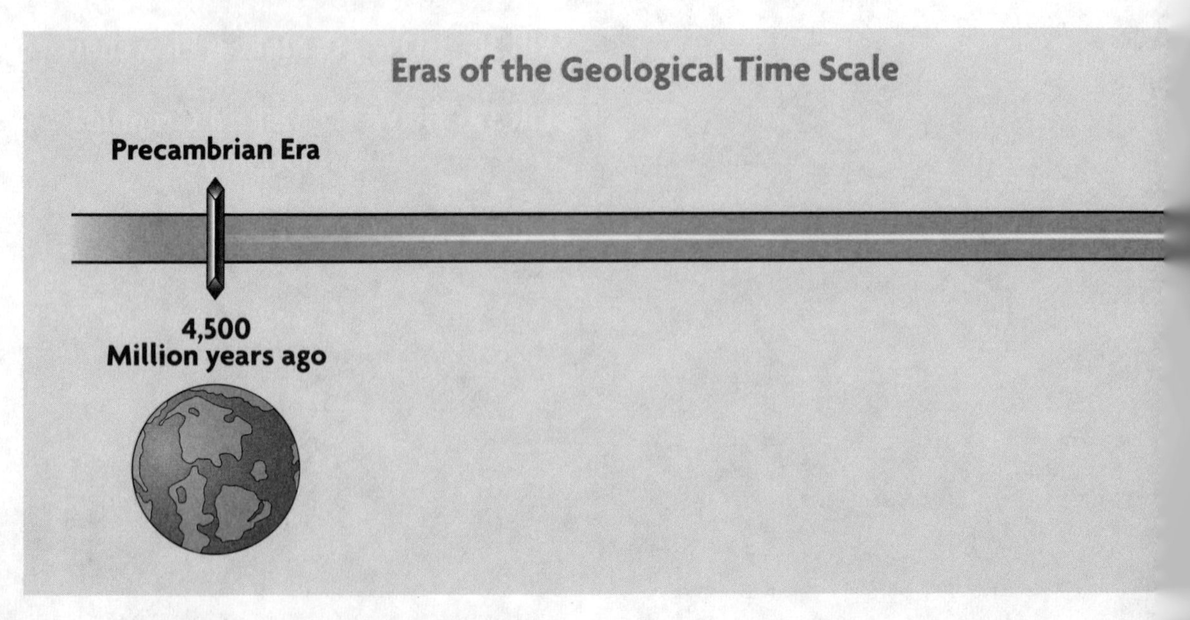

Your birthday is always important to you. Did you ever stop to think that, with each passing year, Earth turns one year older as well? Just as each year of your life brings events and changes, each year brings changes to Earth too.

Earth has existed for billions of years. Over those years, many events have occurred and many changes have taken place. Scientists have divided the history of Earth into four major time periods based on the kinds of life forms that lived on Earth, as well as the physical changes to the surface of Earth. This division of the history of Earth is called the **geological time scale.** The geological time scale starts with Earth's formation more than four billion years ago. Since Earth's beginning, changes to land, water, and climate have occurred. The largest division of geological time is called an **era.** Scientists have further divided the four eras into many shorter amounts of time called **periods.**

Look at the time line below, which shows the four eras of geological time.

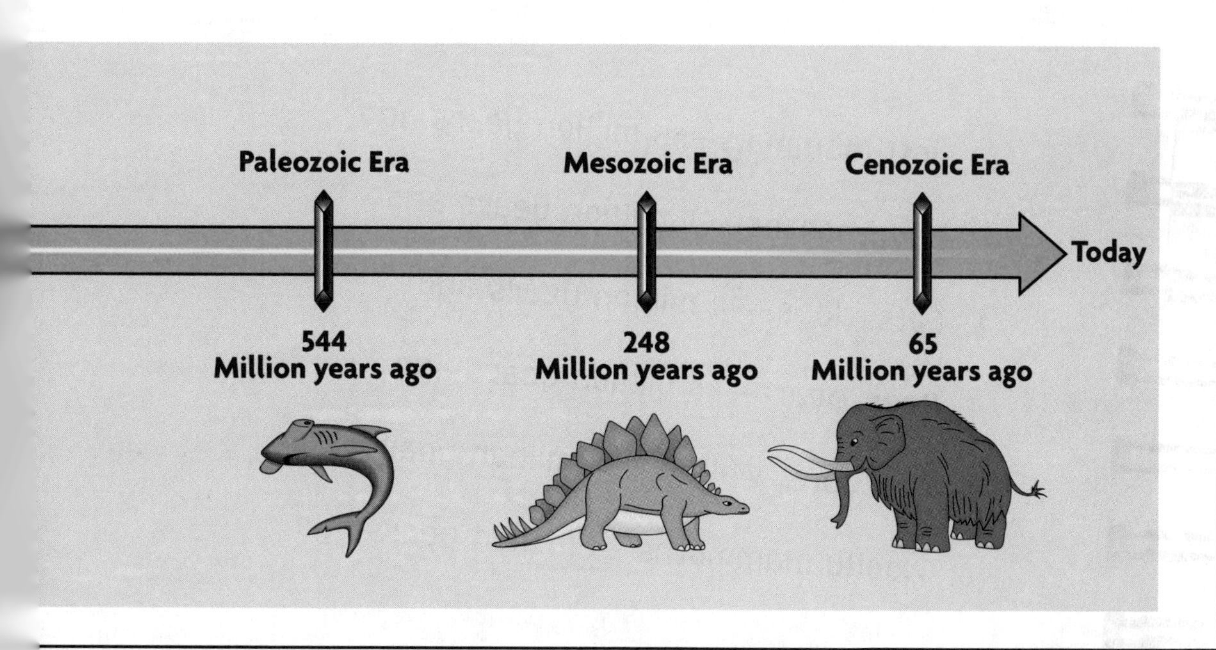

STEPS IN Making a Time Line

You can create a time line by following the steps on these two pages.

1 Decide on Your Subject

Think about what information or events you want to show on your time line. The events should have a common theme.

2 Choose a Title

The title should tell exactly what is being shown on your time line, whether it is a schedule, a series of events, or periods in history. The title should be direct and clear.

First Appearances of Animals During the Cenozoic Era

3 Make a List of Events

List the events or information you want to include in your time line. Use one-word labels or short phrases. Number these events in correct order. You should also write the date or time information of each event.

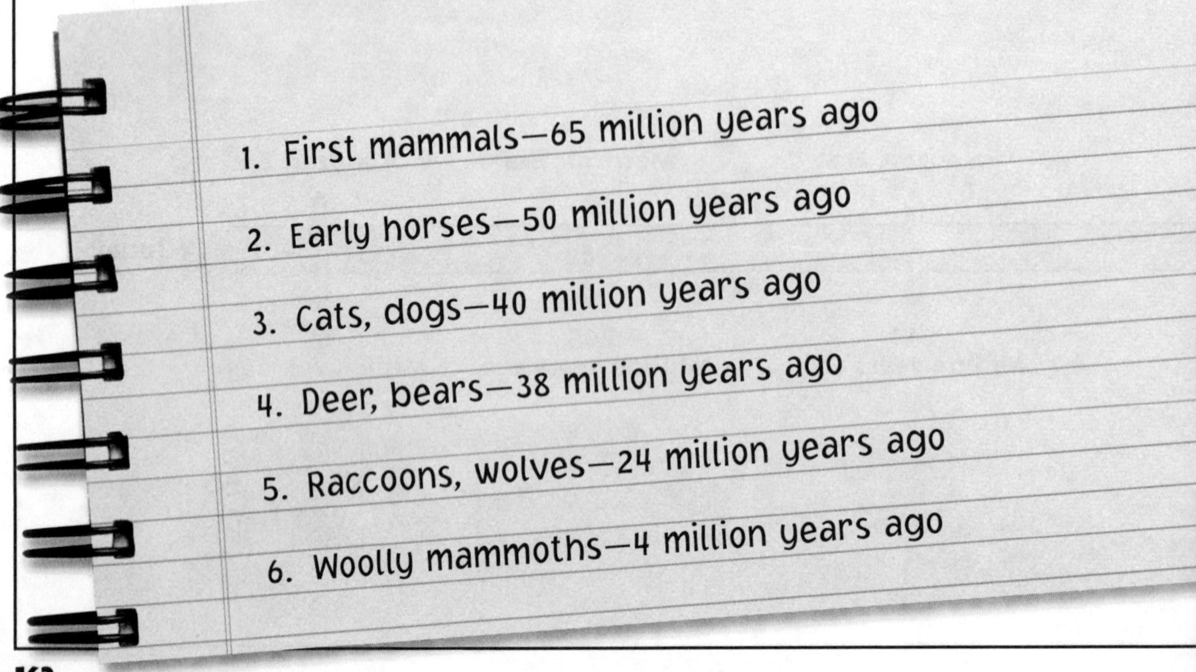

1. First mammals—65 million years ago
2. Early horses—50 million years ago
3. Cats, dogs—40 million years ago
4. Deer, bears—38 million years ago
5. Raccoons, wolves—24 million years ago
6. Woolly mammoths—4 million years ago

4 Draw Your Time Line

- Draw a line from left to right (or from top to bottom, if you wish) on a sheet of paper. Leave space above and below this line to write information.
- Make dots or marks on the line. Each dot stands for the time of an event. The space between each dot on the line indicates the length of time between each event. If the length of time between events is not equal, vary the space between dots to represent the different lengths of time.
- At the appropriate place on the line write the event. Remember to use short phrases.
- Below or next to each dot write a time label. Your time labels might be specific dates, years, or general labels to indicate time, such as names of months.

TIP **A time line will not be easy to read if it is too complicated or filled with too much information. If you have so much information to show that your time line becomes crowded, break the time periods up into several different time lines, each showing a separate block of time.**

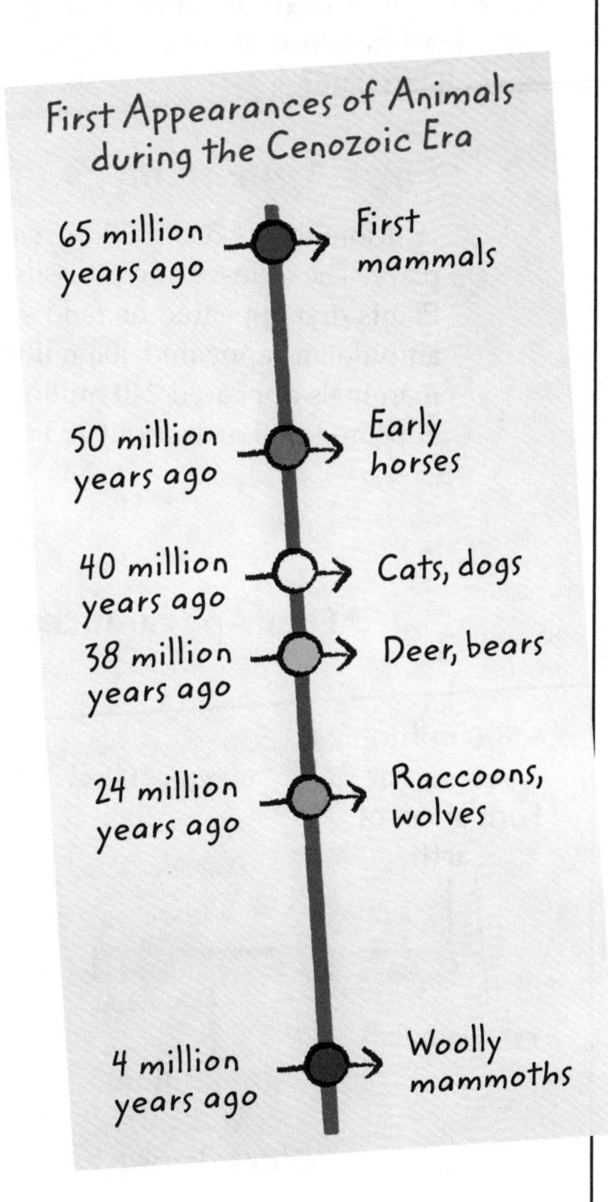

EXAMPLE OF **Making a Time Line**

Read the following information about the first appearances of plant and animal life throughout the eras of geological time. Then find the same information as it is given on the time line. Which form is easier to read and understand?

First Appearances of Plant and Animal Life

More than 4,500 million years ago the formation of Earth took place. The oldest known fossils date back 3,500 million years. Plants first appeared on land 480 million years ago. The first amphibians appeared 408 million years ago, and the first mammals appeared 240 million years ago. Dinosaurs first appeared on Earth about 225 million years ago.

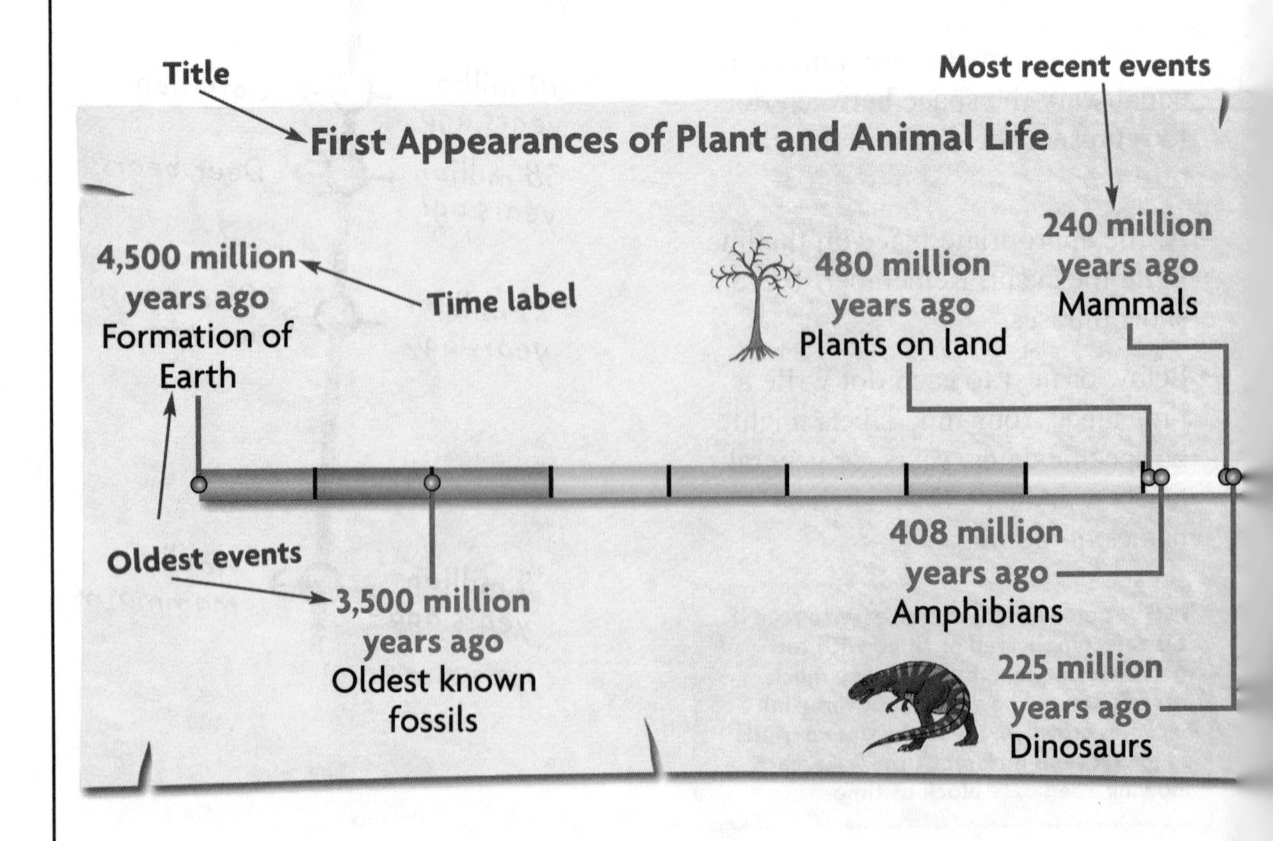

USE THIS SKILL

Make a Time Line

Read the information below about some of the major events during the eras of geological time. The events are ongoing over many years. Use this information to make your own time line, putting the events in order along your time line from the oldest to the most recent dates.

Important Geologic Events of Earth's History
Numbers represent *millions of years ago* (MYA).

250 MYA
Appalachian mountains form

4,600 MYA
Seas form

500 MYA
First fish appear

136 MYA
Dinosaurs die out
Flowering plants appear

680 MYA
Organisms such as jellyfish and clams develop

1.8 MYA
Great Lakes form

430 MYA
Coral reefs form
Air-breathing animals develop

138 MYA
Birds appear

190 MYA
Rocky mountains form

TEST TIP

When taking a test, you may be asked to read information that includes lots of events and dates. You may wish to draw a quick time line on scratch paper to help make the order of the events easier for you to understand.

Skill 33

HOW TO

Read a Table

The World's Strongest Earthquakes

Earth is made up of layers and layers of rock. Rocks deep within Earth are under great pressure. This pressure can cause rocks to slide and break. An earthquake is the result of this movement.

An **earthquake** is the vibrations, or shock waves, caused by movements within Earth. Earthquakes occur every day, but most are so small that people don't even notice them. A **seismograph** is an instrument that measures the strength of earthquakes. Seismographs have been placed in locations around the world. They constantly record movements within Earth. The **magnitude** of an earthquake refers to the strength of the quake. Although most earthquakes can barely be felt, some earthquakes cause massive destruction.

A scale that compares the magnitude of earthquakes is called the **Richter scale.** The table below shows basic measurement categories of the Richter scale.

RICHTER SCALE	
0–2.4	Usually not felt by people
2.5–5.4	Felt by some people; only minor damage
5.5–6.0	Some damage to buildings
6.1–6.9	Danger to people; some destruction of buildings
7.0–7.9	Serious injury and destructio
8.0 and higher	Major devastation; whole communities destroyed

Information and records about earthquakes are often presented in tables. A **table** lists facts and numbers in a way that is easy to read and understand. Tables have headings, or categories, by which the information is organized. They contain rows and columns. Rows go across the page, while columns go down the page.

STEPS IN Reading a Table

Learn how to read tables by following these steps.

1 Identify the Purpose

Find out what kind of data or information is shown by the table. The title of the table should clearly state its purpose.

2 Study the Organization

Read the headings to see how the table is organized. Headings down one side of the chart usually tell what items will be described. The headings across the top of the table tell what kinds of details or facts will be shown about each item.

3 Read the Facts and Figures

A table contains facts and "figures," or numbers, under the headings. Read across from the headings along the side, and read down from the headings along the top. Where the row and column meet, you will find data such as dates, amounts, measurements, or descriptions.

4 Read All Columns and Rows

Make sure to read all the rows and columns of information to get a complete understanding of the table.

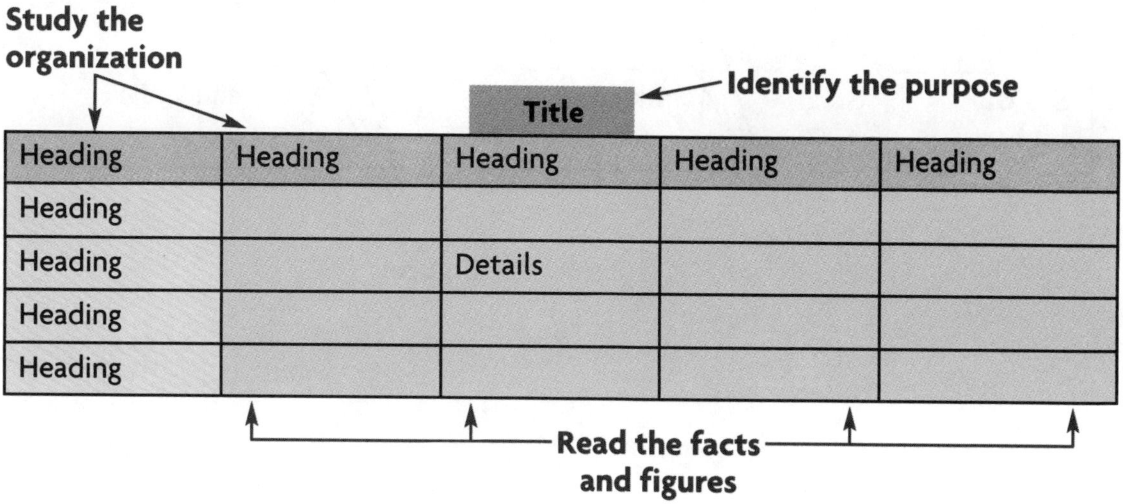

Study the organization

Identify the purpose

Title

Heading	Heading	Heading	Heading	Heading
Heading				
Heading		Details		
Heading				
Heading				

Read the facts and figures

EXAMPLE OF **Reading a Table**

Study the following example of a table.

Headings

Title

Earthquake Facts			
Magnitude	**Class**	**Effects**	**Estimated Number per Year**
0–2.4	Micro	Usually not felt but can be picked up by seismograph	900,000+
2.5–5.4	Minor to moderate	Often felt, minor damage	49,000+
5.5–6.0	Moderate	Slight damage to buildings	800
6.1–6.9	Strong	May cause damage in populated areas	120
7.0–7.9	Major	Serious damage	18
8.0 or higher	Great	Total destruction	Between 1 and 4 every few years

Facts

Figures

USE THIS SKILL

Read a Table

Study the table on this page. Use the information in the table to answer the questions below.

Strong Earthquakes			
Year	**Location**	**Richter Scale**	**Number of Deaths**
1556	Shensi, China	Unknown	830,000
1906	San Francisco, California	8.3	1,500
1920	Kansu Province, China	8.5	180,000
1939	Concepcion, Chile	8.3	30,000
1964	Prince William Sound, Alaska	8.5	131
1970	Peru	7.8	66,800
1976	Tangshan, China	7.6	240,000
1985	Mexico City, Mexico	8.1	9,500
1990	Iran	7.7	50,000
1993	Guam	8.1	none
1994	Northridge, California	6.7	61
1995	Kobe, Japan	6.9	5,378

1. According to the table, in what year did the earthquake occur in Shensi, China?
2. Which two earthquakes in the table had the highest magnitudes according to the Richter scale?
3. What was the magnitude of the earthquake in Mexico City in 1985?
4. Where did a strong earthquake happen in the year 1995?

TEST TIP

Some tests provide tables of information for you to read and understand. Make sure that you read straight across rows and straight down columns. Use your finger or the edge of a sheet of paper to help you follow the information correctly.

Skill 34

HOW TO

Make a Bar Graph

Choosing Healthful Foods

Do you ever have trouble deciding what to eat? The cupboards and refrigerator in your home are probably full of tempting choices. How do you choose?

One way to choose foods is for their nutritional value. Without the proper nutrients, your body may have difficulty growing or fighting off disease. It is important to eat healthful foods at snack time and at mealtime.

What does healthful mean? In general, it means eating a large variety of foods as recommended on the Food Guide Pyramid. This way you will be getting all the vitamins, minerals, proteins, carbohydrates, and fats that your body needs. Whenever possible, it is also important to eat fresh foods. Fresh foods are full of nutrients, without the added sugar, salt, and fat found in prepared packaged foods.

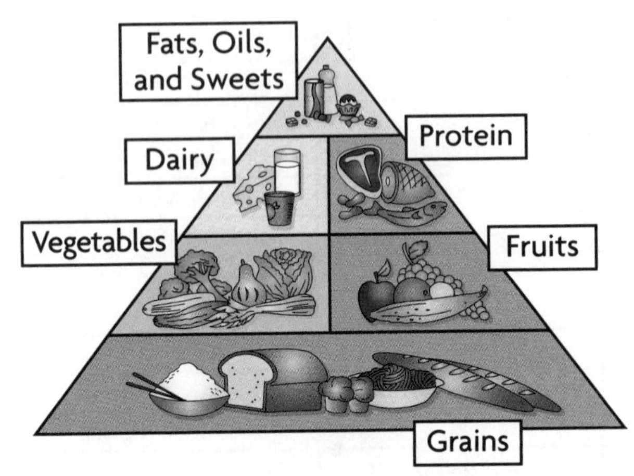

The Food Guide Pyramid

Consuming too much salt or saturated fat can lead to heart and circulatory problems. Too much sugar can cause tooth decay and fills you up with calories that have no nutritional value.

As you plan a healthful diet, you may find it helpful to use graphs to display the nutritional values of various foods. A **bar graph** uses bars, lines, and numbers to compare two or more things. A bar graph can show information at a glance so that you can make choices more easily.

STEPS IN Making a Bar Graph

1 Collect Information

Before making a bar graph, you need to collect information that shows the differences between your subjects. Give your graph a title that tells what information will be shown.

2 Set up the Graph

Bar graphs are drawn either up and down or across. To make a bar graph that goes up and down, or vertically, begin by drawing a large L. The groups of items to compare will fall along the bottom line. The amounts you are comparing will fall along the line going up. The bars on your graph will begin at the bottom and go up.

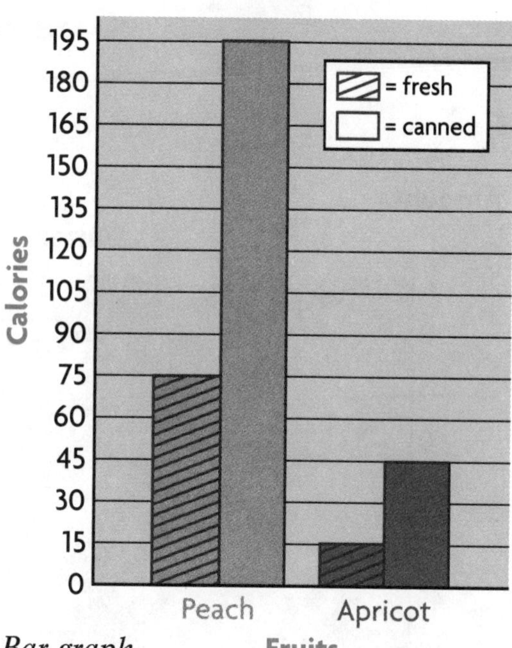

Bar graph

3 Draw the Graph

Write a zero at the outside bottom left corner of your graph. Then find the largest number you need to record and place it at the top of the same line. Decide at what intervals you want to mark the rest of the line. Use a ruler to mark the line and label each mark with a number. List the groups of items you are comparing along the bottom. Some bar graphs have two bars that represent the two items being compared. To make these "double" bar graphs, draw a bar for each item you wish to compare. Use two different colors or patterns to represent the two items in each group. Use a ruler to make the top of your bars the same height as the correct amount on the line. Then draw a small box, or legend, in an empty area of the graph. Inside the box, write what each color or pattern represents.

EXAMPLE OF Making a Bar Graph

Look at the bar graph below. The bars show the amounts of two important minerals in different vegetables. Both minerals are measured in milligrams.

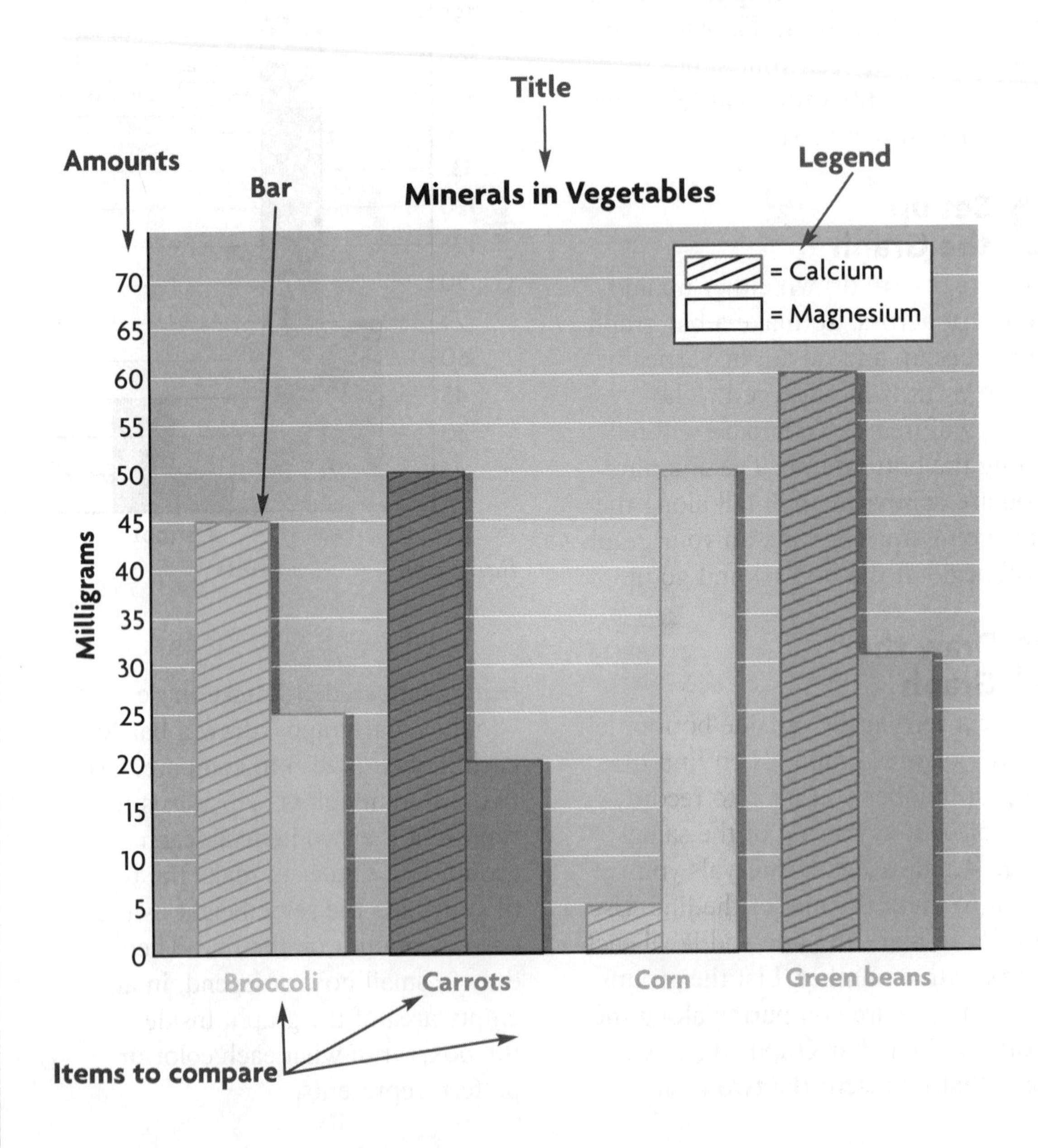

USE THIS SKILL

Make a Bar Graph

Read the passage below. Follow the directions to make a bar graph.

Choosing Healthful Main Dishes

Mrs. Rodriguez wants her family to eat nutritious meals. She knows the importance of selecting main dishes that are high in protein but low in fat. Mrs. Rodriguez always gathers information so that she can make informed decisions on which dishes to serve to her family. Below are five main dishes and the number of grams of protein and fat in each serving. Make a bar graph that will help Mrs. Rodriguez examine the information more easily.

	Protein	Fat
Beef stew	25 grams	13 grams
Spinach quiche	8 grams	21 grams
Turkey sandwich	14 grams	6 grams
Baked cod	28 grams	3 grams
Chicken pot pie	11 grams	29 grams

TEST TIP

You may be asked to read a bar graph on a test. First read the title, all the labels, and the legend so that you understand what information is being shown. Then look carefully to see where each bar ends. If a bar falls between two labeled amounts, you will need to estimate its value.

Skill 35

HOW TO

Read a Line Graph

Predator and Prey Populations

In the wild, animals have to find or hunt their own food. Some animals eat plants or plant products, while others must eat other animals to survive.

An animal that hunts and depends on another animal for its food is called a **predator.** The animal that it hunts is its **prey.** Predators and prey have an important relationship in the environment that they share. This relationship determines the size of populations of both predators and prey. For example, the fisher, a relative of the weasel, is one of the few animals able to kill porcupines. In Wisconsin, the fisher population almost disappeared because of human hunting. As a result, porcupines became a serious pest. After fishers were reintroduced to the area, porcupines were no longer a problem because the fishers killed many of them.

The fisher is a predator.

To learn about changes in predator and prey populations, scientists collect data to help them study patterns of change. A **line graph** is a tool that uses grids, lines, and dots to show how amounts of something change over time. Like most graphs, line graphs provide a lot of information in a small amount of space. It is important to know how to read a line graph so you can see relationships and trends. Line graphs usually appear on squares that form a grid. Plotted inside the graph are points and lines. The steps on the next page show how to read a line graph.

STEPS IN Reading a Line Graph

1 Understand the Purpose

First read a line graph's title to learn the specific purpose of the graph. For example, a graph titled *Number of Polar Bears on Mt. Snow* tells you that the graph focuses only on the number of polar bears counted on Mount Snow. Next look for the axis lines. The **horizontal axis** is the line that runs along the bottom of the graph. It shows the passage of time, and it's usually labeled with time units that increase as they go toward the right. The **vertical axis** is the line that begins at zero and travels up the side of the graph. The subject or subjects are labeled here. As this line goes higher, the numbers get larger.

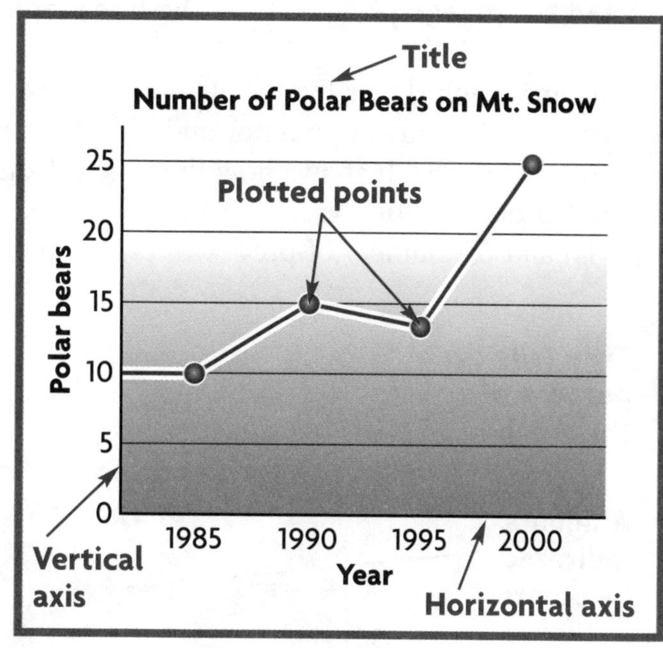

2 Read the Data

Follow the line and read the amounts indicated by the dots, or points, on the graph. For example, if you want to find out how many polar bears were counted in a particular year, start at the *year* with a finger and travel up to the dot or plotted point. Use this same finger to travel in a straight line to the *number* figure on the left.

3 Find Trends and Relationships

The lines connecting the points inside a graph show changes in amounts over time. A line on a graph may go down, showing a trend such as a decreasing population. The graph may also show two or more lines, to show that two or more subjects are being compared. The patterns made by these lines can show relationships, such as how predator populations are affected by prey populations. By showing patterns, line graphs help readers make predictions about what might happen next.

EXAMPLE OF Reading a Line Graph

Read the paragraph and study the line graph below.

This line graph shows the results of an eight-year study of predator and prey populations. It shows how the number of predators (great horned owls) and the number of prey (white-footed mice) changed over time in Spring Valley, a large nature preserve. Notice that there are trends and relationships between the two populations.

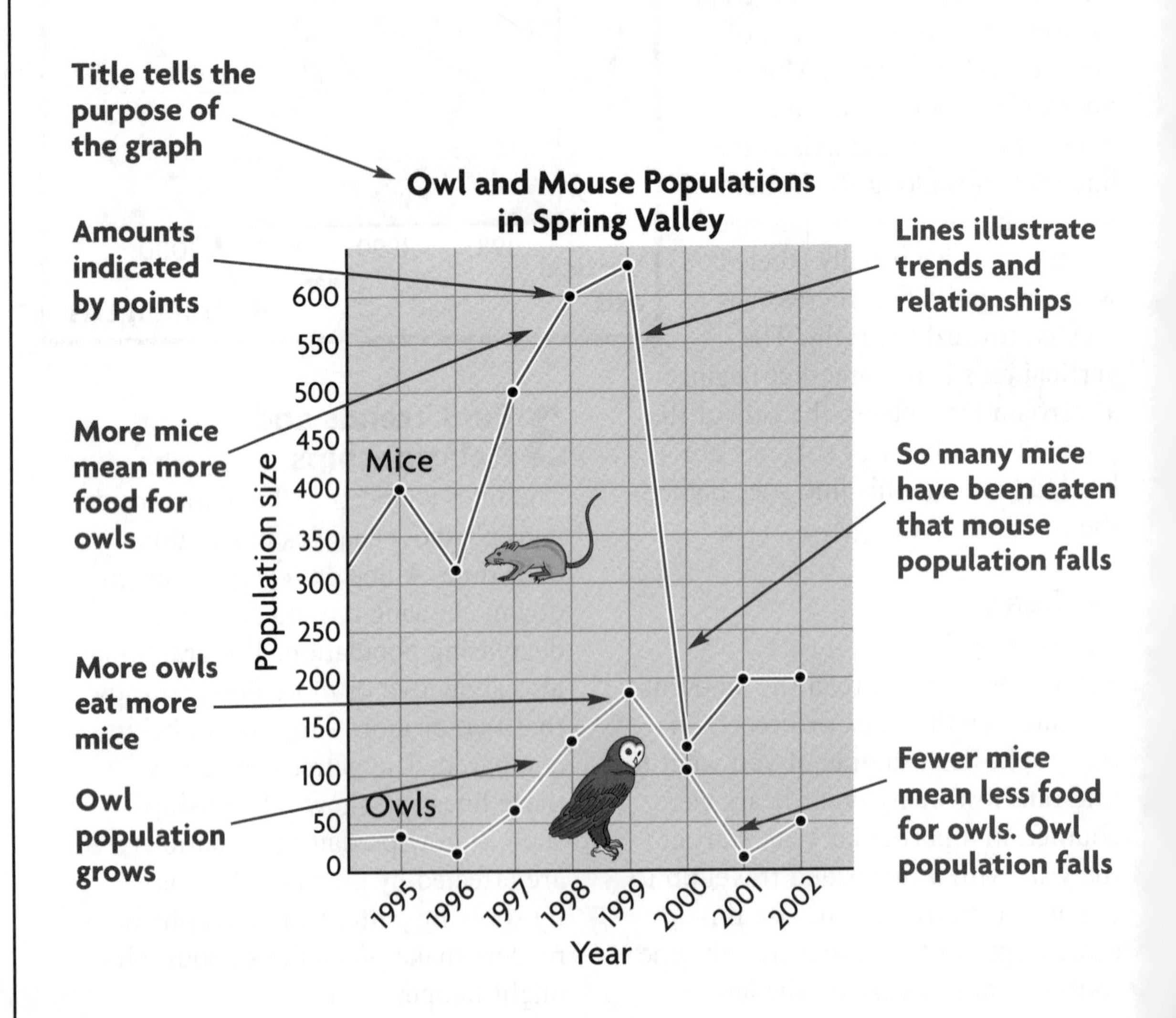

USE THIS SKILL

Read a Line Graph

Read the line graph and answer the questions below.

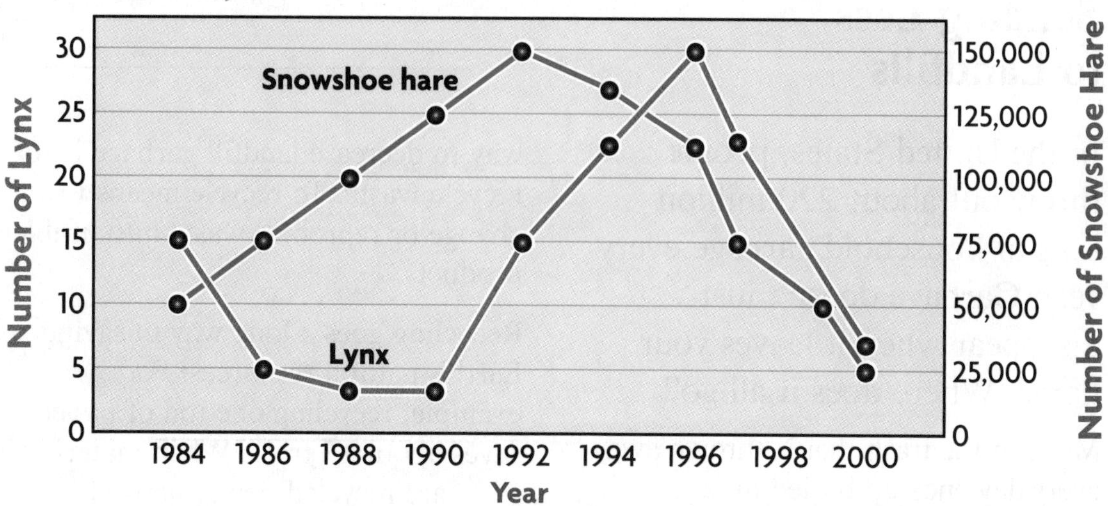

1. What are the labels on the two vertical axes?
2. What is the label on the horizontal axis?
3. Which years had the fewest lynx?
4. When the snowshoe hare population increases, what happens to the lynx population?
5. When the snowshoe hare population decreases, what happens to the lynx population?
6. Describe the trend between the snowshoe hare population and the lynx population shown on this line graph.

TEST TIP	When you answer questions about a line graph on a test, quickly look over the graph to get started. Look back at the graph to answer each question. Don't try to memorize the information shown on the graph.

Skill 36

HOW TO

Read a Circle Graph

Sending Less to Landfills

In the United States, people throw out about 220 million tons of household garbage every year. Garbage doesn't just disappear when it leaves your home. Where does it all go?

Much of the trash that is thrown away every day ends up buried in landfills. A **landfill** is a place set aside for the dumping of garbage. The United States has more than 9,000 landfills. The big problem is that landfills are filling up, and the areas where landfills can be made are limited. People don't want to live next to landfills, so they usually protest the development of new landfills. Landfills often smell bad, they can pollute the air and water, and they can interfere with the balance of local ecosystems.

Reduce!
Reuse!
Recycle!

The amount of garbage ending up in landfills needs to be reduced. One way to decrease landfill garbage is to recycle waste. To **recycle** means to change or reprocess waste into usable products.

Recycling goes a long way in saving Earth's natural resources. For example, recycling one ton of paper saves about 17 trees. When materials are recycled, fewer natural resources are used and less waste is buried in landfills.

If you wanted to start a recycling program for your home or school, you could begin by keeping track of the kinds of garbage you throw away. Making a chart or graph could help you organize this information. A **circle graph,** sometimes called a pie graph, is a way to show how parts of a whole are divided up. The whole circle represents 100 percent, or all, of something. The circle is divided into sections. Each section is a part, or percentage, of the whole circle.

STEPS IN Reading a Circle Graph

Follow these steps to learn how to read a circle graph.

1 Find the Title

Read the title to find out the subject of the circle graph. The title of a graph may include words and numbers.

2 Study the Labels

Labels describe what each section represents. The labels will also give numbers or percentages of the whole amount being shown on the circle graph.

3 Compare the Sections

Notice the different sizes of the sections. Compare the size of each section to the sizes of the other sections. The smallest section of the circle graph represents the smallest part of the whole amount. The largest section of the circle graph represents the largest part of the whole amount.

TIP All numbers in the sections of a circle graph must equal the total number of the whole graph when added together.

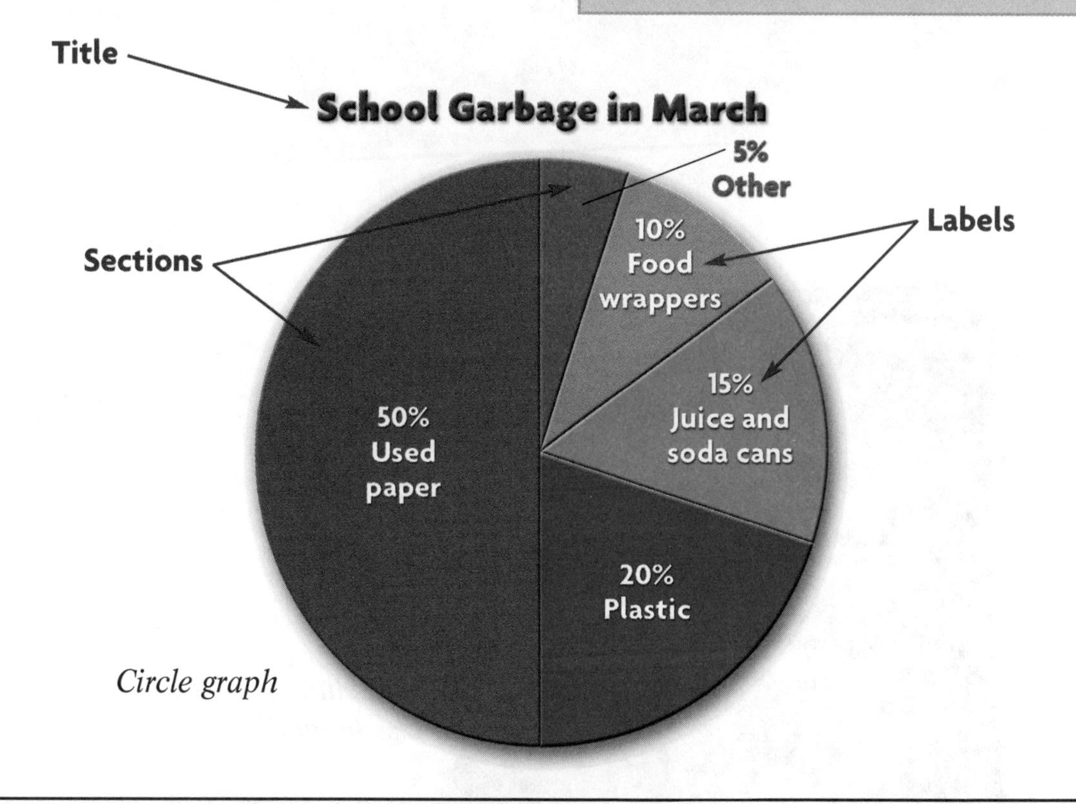

Circle graph

EXAMPLE OF **Reading a Circle Graph**

The United States creates millions of tons of garbage every year. Read the following circle graph that shows the sources of different types of waste.

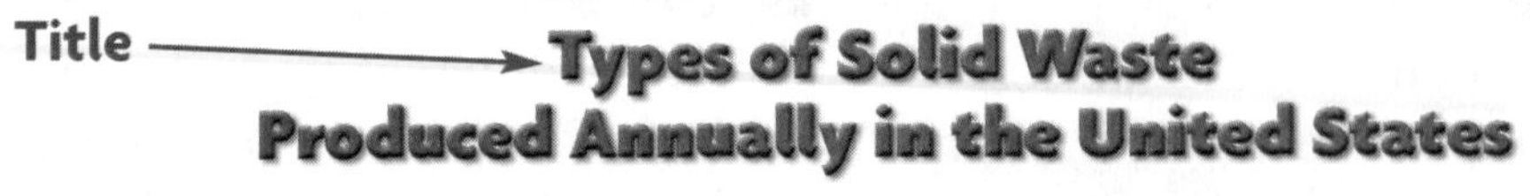

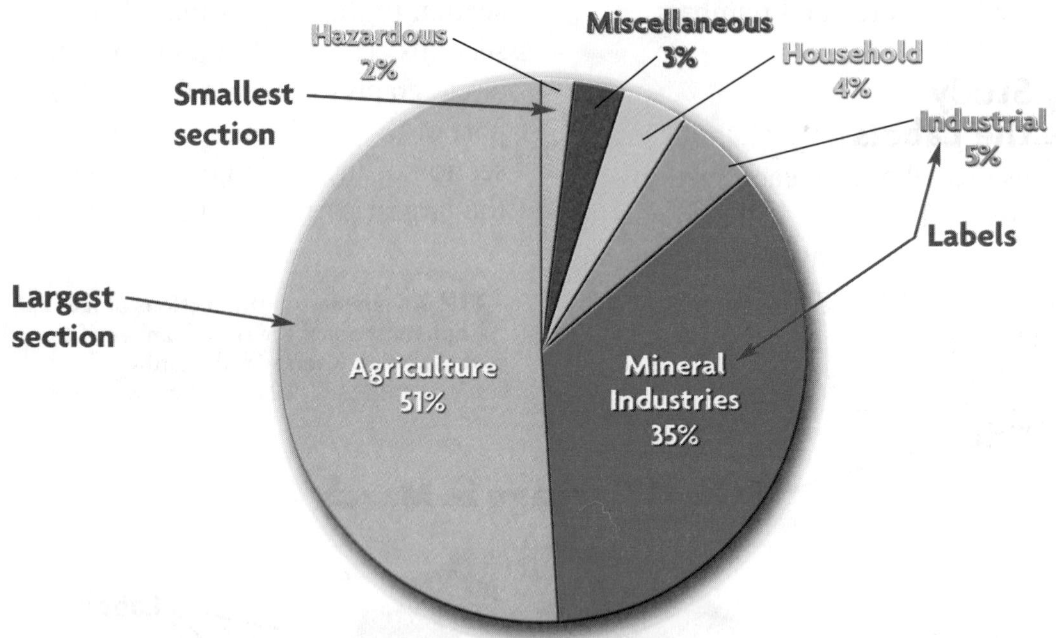

Recycling helps send less to landfills

USE THIS SKILL

Read a Circle Graph

Study the circle graph below, which shows the weekly garbage produced by a typical American family of three people. Then answer the questions.

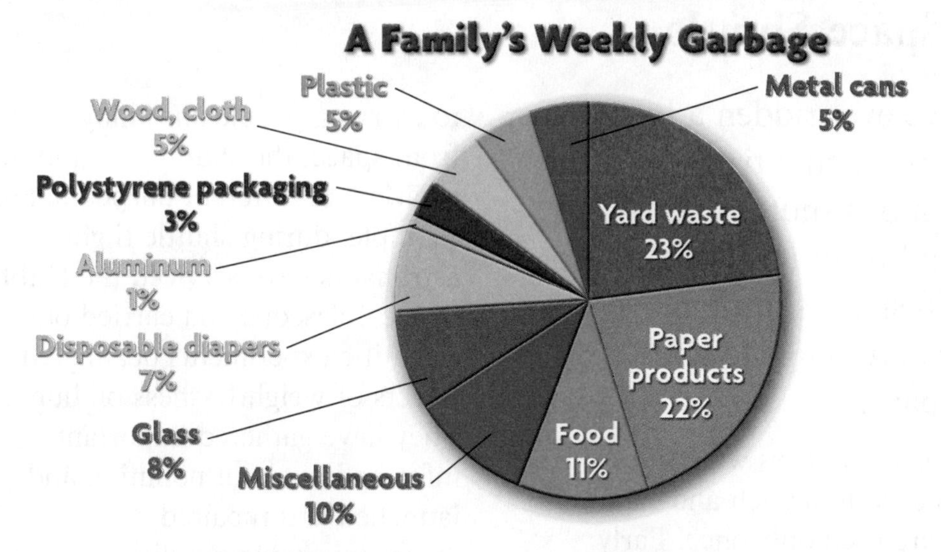

1. What subject does the whole circle graph represent?
2. What two kinds of garbage make up the largest amount of waste?
3. What percentage of garbage is glass?
4. What kind of garbage makes up 1 percent of the graph?
5. What percentage of garbage would be reduced if the following kinds of garbage were recycled?
   - paper
   - yard waste/grass clippings
   - glass
   - plastic
   - metal cans
   - aluminum
   - polystyrene packaging

TEST TIP

When answering test questions about a circle graph, read the questions carefully to be sure you are looking at the section of the graph to which the question is referring. Check your answers by looking at the graph again.

Skill 37

HOW TO

Read a Diagram

The Space Shuttle

If you've ever ridden a bus, then you've been on a type of shuttle. Just as a bus shuttles people back and forth to places on Earth, a space shuttle moves people and things to and from space.

Before space shuttles were in use, vehicles used to launch and explore space were used only once. Early space vehicles burned off in layers upon returning to Earth's atmosphere. They could not be used again. In an effort to save time, money, and resources, NASA leaders created a reusable spacecraft, known as the **space shuttle.**

The first U.S. space shuttle, *Columbia,* was launched in April 1981. Since then, four more have taken flight—*Challenger, Discovery, Atlantis,* and *Endeavour.* In addition to carrying people and things to and from space, the shuttle program has a variety of different purposes. For example, during shuttle flights, astronauts have serviced the Hubble Space Telescope and carried out scientific experiments dealing with the effects of weightlessness on humans. They have gathered important information about pollution and launched and repaired communications satellites.

So what does the inside of a space shuttle look like? What are the functions of a shuttle's different parts? This is where a diagram can help. A **diagram** is a sketch, picture, plan, or model that helps explain or illustrate information. Diagrams are often included in books or articles to help you get a better understanding of what you read.

There are many different kinds of diagrams. For example, maps show places, such as the boundaries and cities in a nation or state. Line diagrams use words and symbols to show how things are related to one another. A **picture diagram** is a drawing or photo of a subject with labels that tell more about what is being shown.

Look at the picture diagram below to see the three parts of a space shuttle before liftoff.

Parts of a Space Shuttle

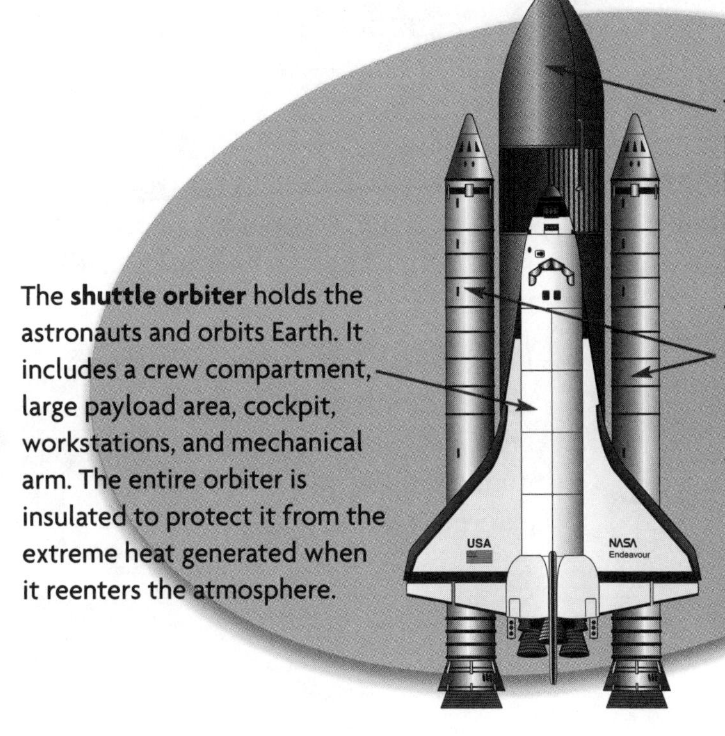

The **external tank** contains liquid hydrogen and liquid oxygen used by the space shuttle's main engines. During liftoff, the tank breaks into pieces. It is not used again.

The **solid rocket boosters** (SRBs) provide the space shuttle with most of the thrust, or force, needed during liftoff and the first part of the ascent.

The **shuttle orbiter** holds the astronauts and orbits Earth. It includes a crew compartment, large payload area, cockpit, workstations, and mechanical arm. The entire orbiter is insulated to protect it from the extreme heat generated when it reenters the atmosphere.

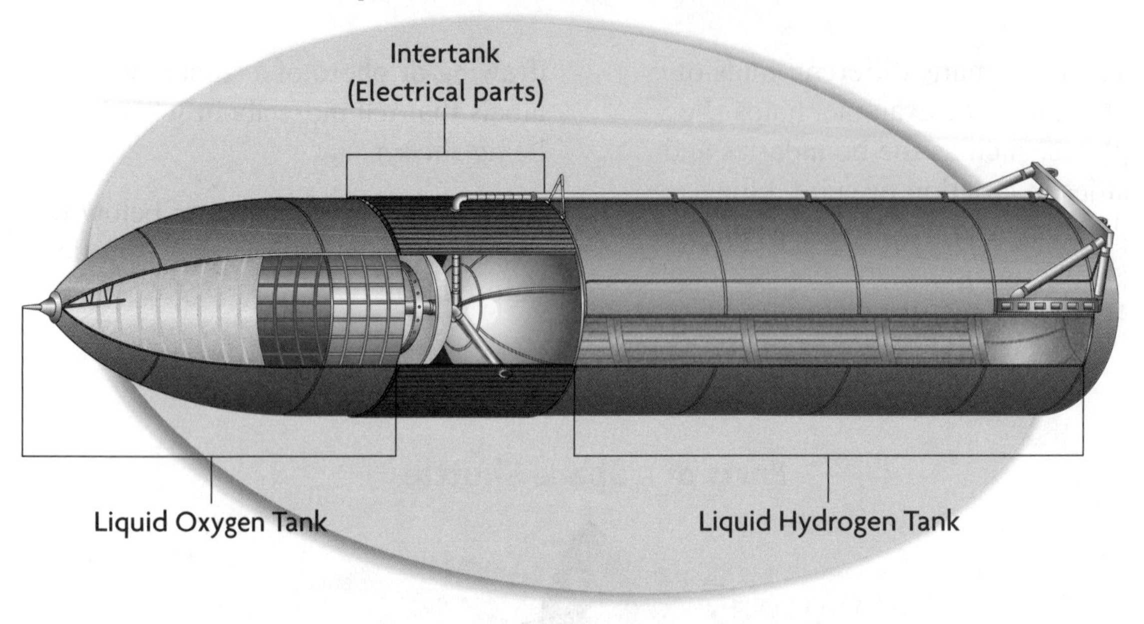

Some diagrams appear to be three-dimensional—they show not only height and width but also the depth of a subject. Often, such diagrams are "cut" down the middle, allowing the reader to "see" inside the subject. This is called a cross section. Some diagrams give you several different views of the subject, such as one view from the top down and then a second view from one side. When you see a few different views of a subject, you get a better idea of what it looks like in real life.

Look at the cross-section diagram above. Think about how getting a glimpse inside helps you better understand how a space shuttle's external tank looks.

STEPS IN Reading a Diagram

Follow these steps to learn how to read a diagram.

1 Read the Title

The title of a diagram tells you what the diagram is about. Read the title first before you study the diagram itself.

2 Read Labels

Look at the labels to see what each part of a diagram represents. Pointer lines tell which labels go with which parts. In groups of diagrams that show different views of the same thing, labels that appear in one of the diagrams may not appear in the others. Or, one label may point to the same part in all of the different diagrams. This allows you to see how the same part looks from different angles.

3 Find the Path

Some picture diagrams should be read from top to bottom—from the title down. Other diagrams should be read from left to right, or bottom to top, or in a circle, depending on the content. When you study a diagram, figure out if there is a certain place where you should begin reading. Follow arrows, if there are any, to find the diagram's path.

4 Study the Information Shown

Study the different parts of a diagram to get an idea of how they are related to one another. When you have read and understand a section, move on to the next one. Make sure you read the entire diagram, not just one part of it.

TIP **To get an idea of how the pieces fit together, think of cross-section diagrams as models you could put together and then pull apart.**

MODEL AIRPLANE GLUE

EXAMPLE OF Reading a Diagram

Look at the diagram below to see some of the parts and functions of a space shuttle orbiter.

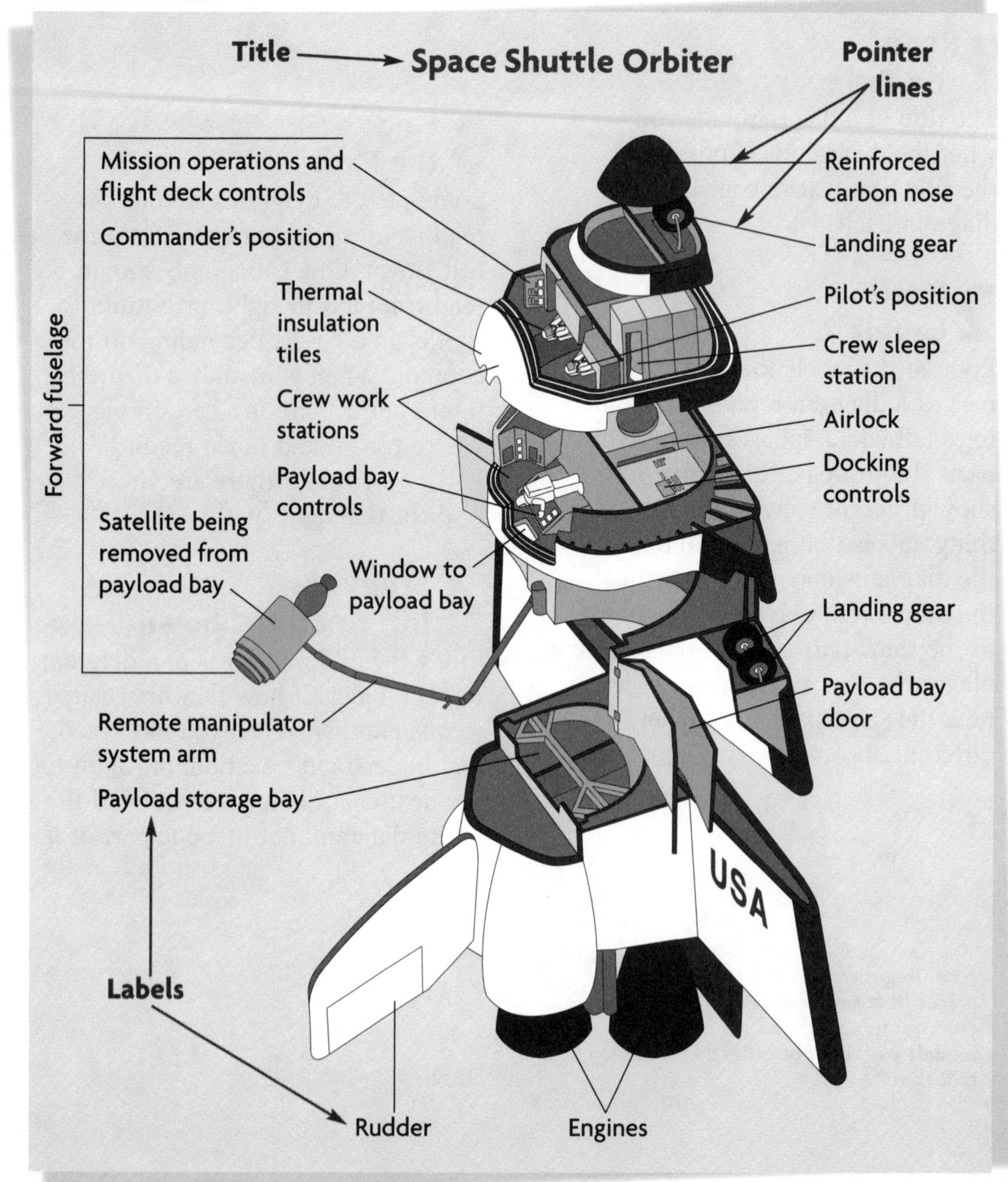

USE THIS SKILL

Read a Diagram

Use the diagram on page 186 to answer the questions below.

1. Which part of the space shuttle is shown in this diagram?
2. In what part of the orbiter does the crew live and work?
3. Where is the sleep station located?
4. Where are the engines located?
5. Where is the payload storage bay?
6. How can the crew see what is inside the payload bay?
7. How many crew members are visible in this diagram?
8. The orbiter lands on the ground like an airplane. How many sets of landing gear can you see in the diagram?
9. Name three types of controls located in the forward fuselage.
10. What is used to move satellites from the payload storage bay?

TEST TIP	Some tests may ask you to answer questions about information presented in a diagram. Read the entire diagram carefully before answering any of the test questions. Look back at the diagram as you answer each question.

Skill 38

HOW TO

Make a Flowchart

A Network of Nerves

When you prick your finger, your hand pulls away. When you bite into a lemon wedge, your mouth puckers. When you breathe in dust, you sneeze. You may not stop to think about these automatic responses. But how and why do they happen?

The **central nervous system,** which includes the brain and spinal cord, allows sense responses to occur. Everything you sense by touch, sight, smell, taste, or sound is communicated to your brain through a network of nerve fibers. A stimulus, such as a hot object, activates sensory nerve cells in the skin.

A message then passes from nerve cell to nerve cell, until it reaches first the spinal cord and then the brain. An impulse from the spinal cord travels down the arm across motor nerve cells. When this impulse reaches the muscle fiber of the arm, the muscle contracts. The hand pulls away from the hot object. The brain also interprets the message from nerve cells, and you sense pain from touching the hot object. This process occurs in a fraction of a second. If it did not, your body would suffer great damage. The nervous system warns the body of potential danger.

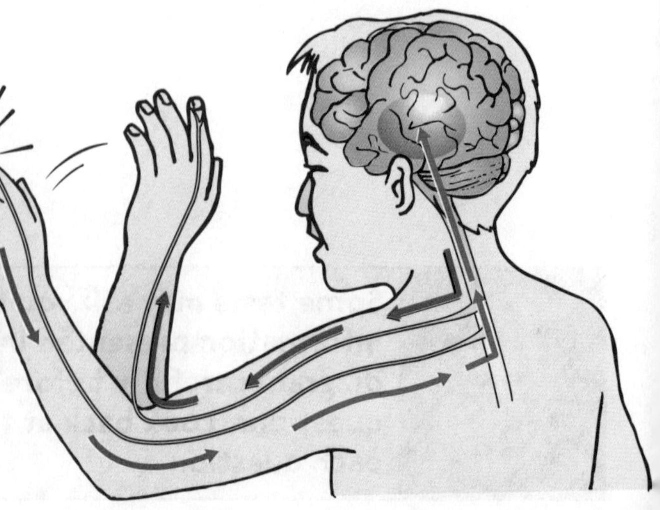

The nerve response to a stimulus takes many different steps. A **flowchart** shows all the steps of a process in order. Arrows point out which step comes after another. If you follow the arrows, you can follow the order of the steps in the process. Look at the flowchart below.

Nerve Response to a Stimulus

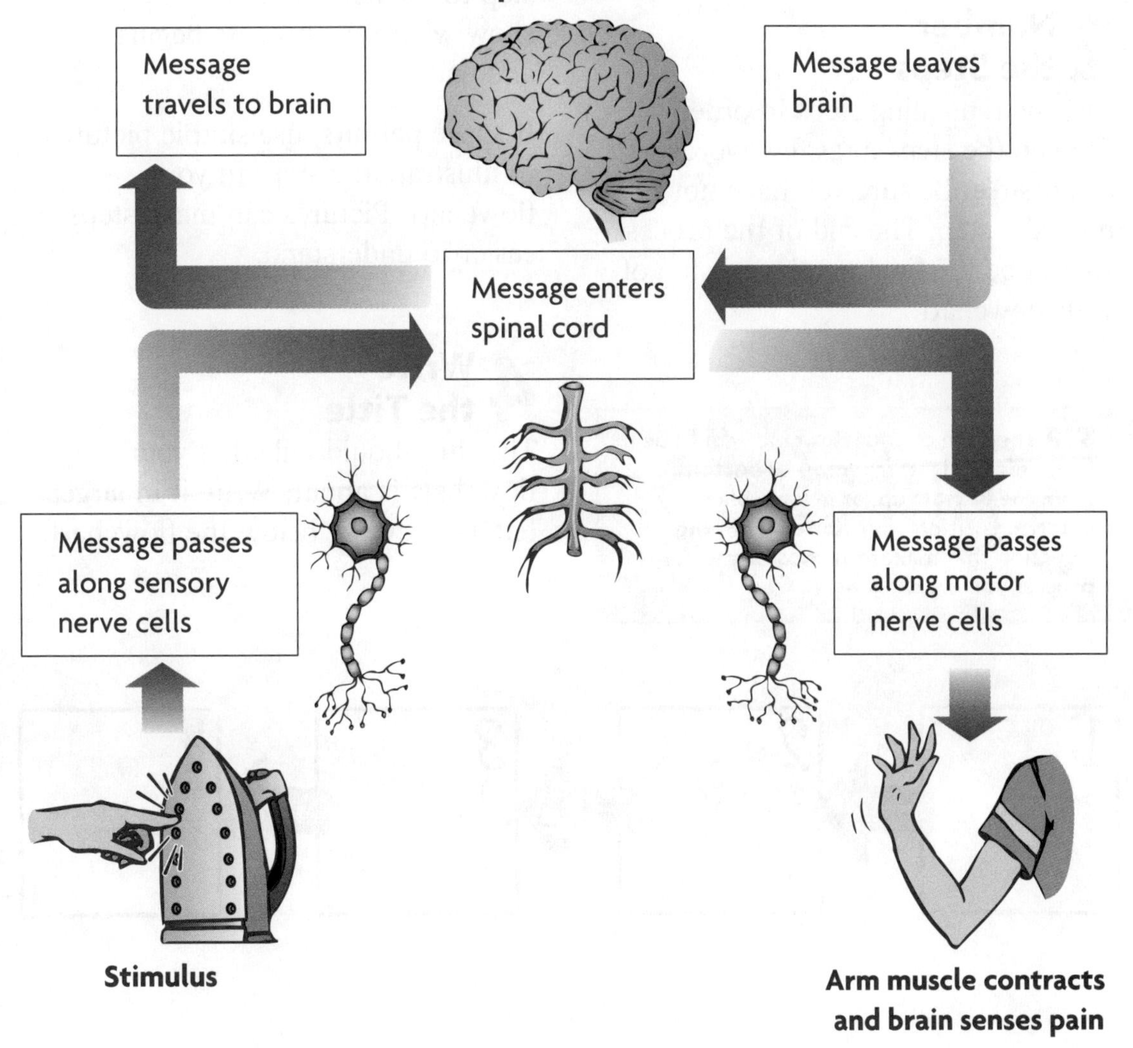
Message travels to brain
Message leaves brain
Message enters spinal cord
Message passes along sensory nerve cells
Message passes along motor nerve cells
Stimulus
Arm muscle contracts and brain senses pain

STEPS IN Making a Flowchart

Follow these steps to make your own flowchart showing a process.

1 Find the Beginning

Find the first step of the process. List the first step in a few words.

2 Number the Steps

List the remaining steps in order. Explain the steps in as few words as possible. Be sure you have not missed a step. The end of the process should be the final numbered step of your flowchart.

TIP **The path of your flowchart could be from left to right, from top to bottom, from the bottom up, or in a circular pattern. Position your flowchart in the direction that makes the most sense for the process you are showing.**

3 Draw Boxes and Arrows

Draw a box around each step and connect each box with an arrow. Arrows show movement from one step to the next. They also help show where the process begins and ends.

If space permits, use simple pictures to illustrate the steps in your flowchart. Pictures can make steps easier to understand.

4 Write the Title

The title should tell what your flowchart is about. Write it in larger letters above or below the flowchart.

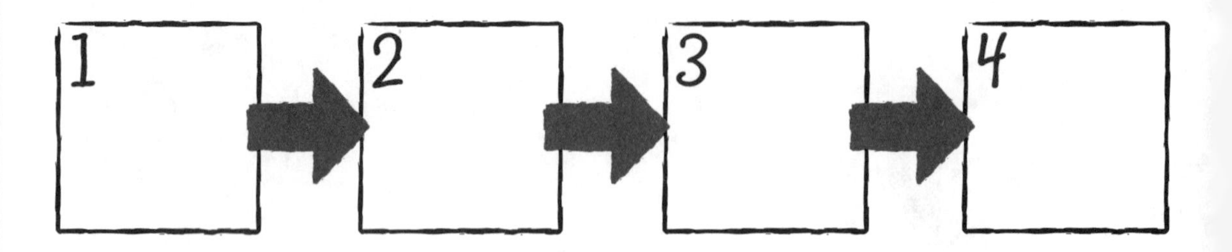

Your Sense of Touch

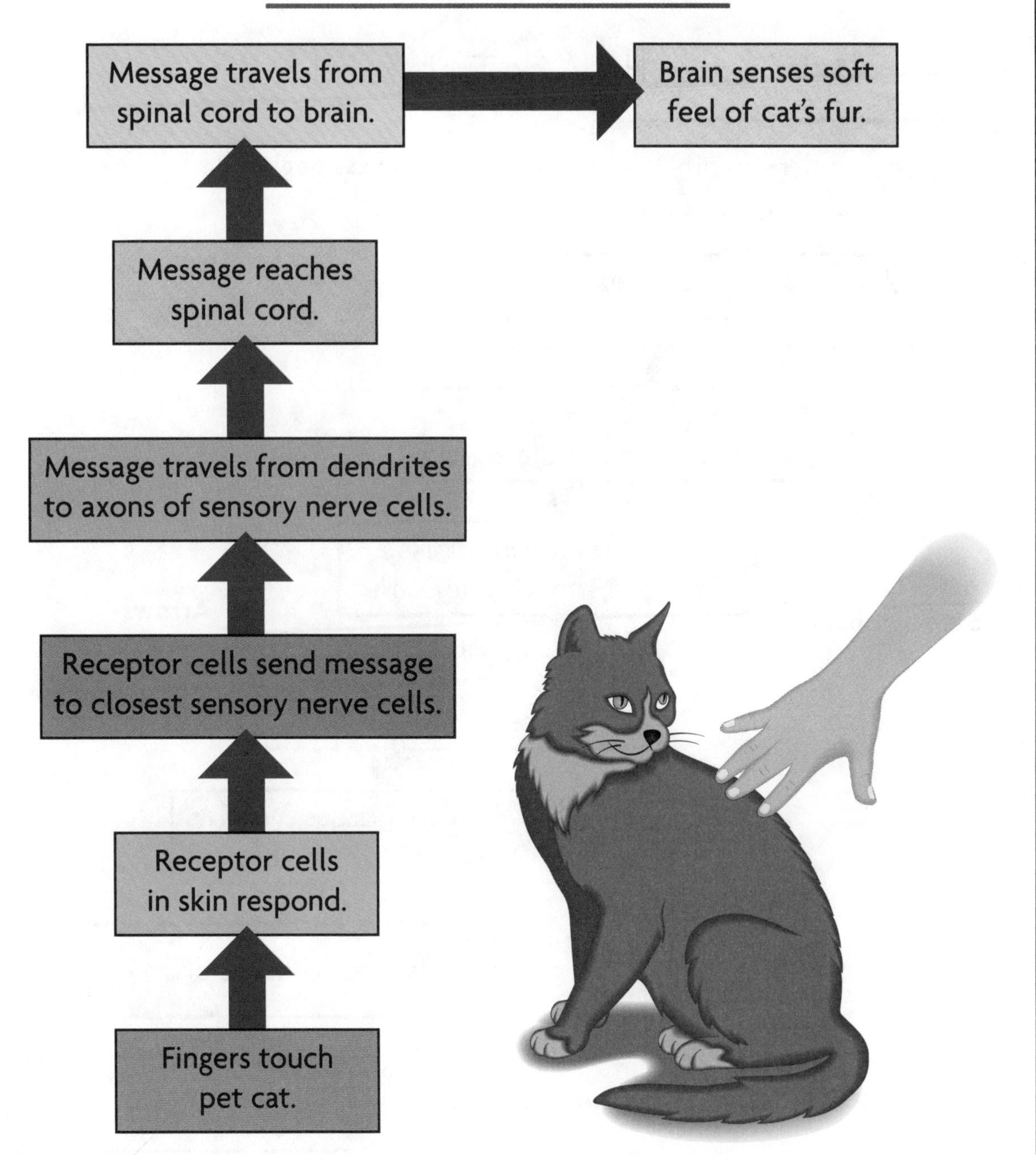

Message travels from spinal cord to brain.
Brain senses soft feel of cat's fur.
Message reaches spinal cord.
Message travels from dendrites to axons of sensory nerve cells.
Receptor cells send message to closest sensory nerve cells.
Receptor cells in skin respond.
Fingers touch pet cat.

EXAMPLE OF Making a Flowchart

See how one student made a flowchart to explain the steps in the taste sense process.

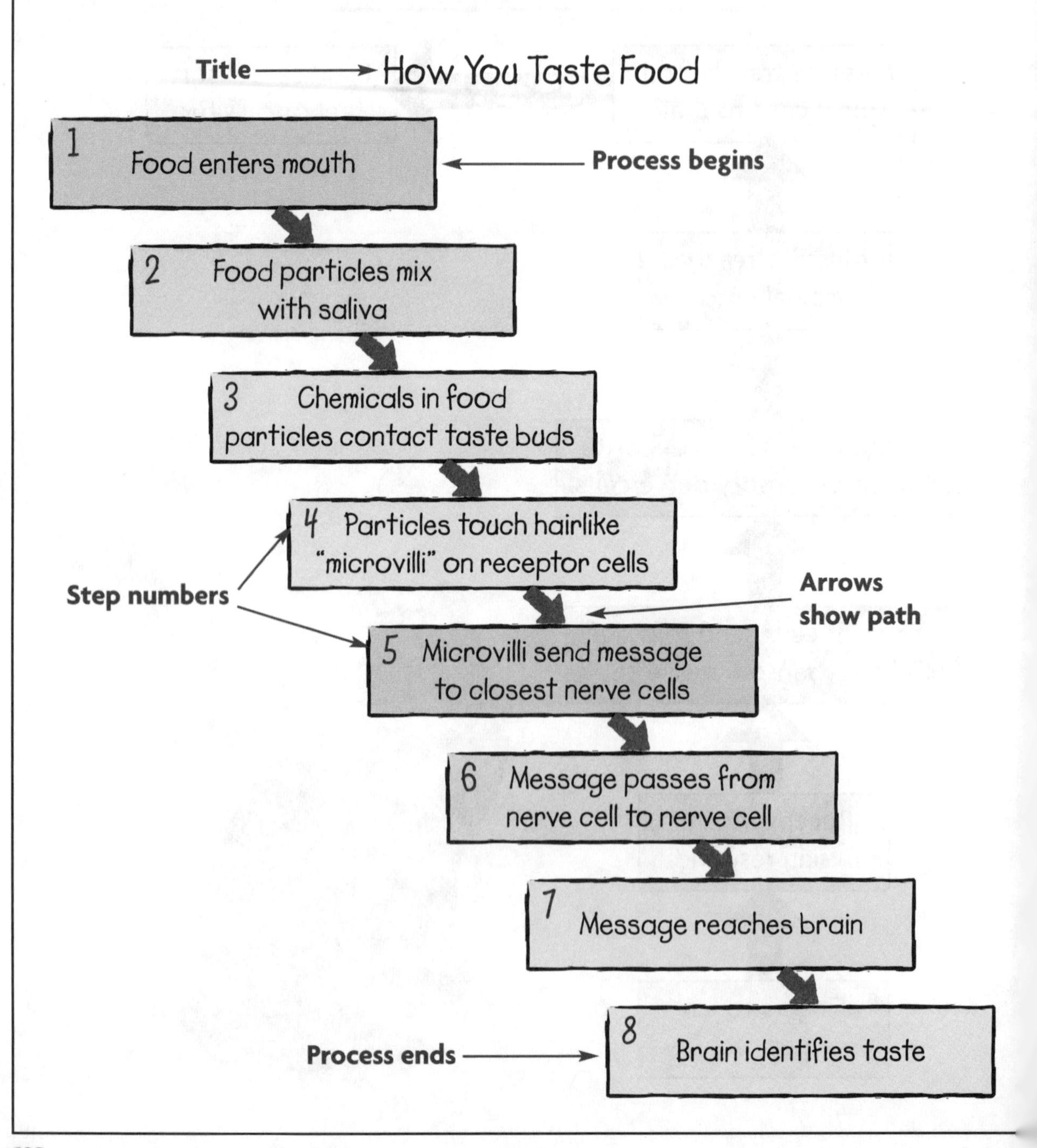

Title
How You Taste Food
1 Food enters mouth
Process begins
2 Food particles mix with saliva
3 Chemicals in food particles contact taste buds
4 Particles touch hairlike "microvilli" on receptor cells
Step numbers
Arrows show path
5 Microvilli send message to closest nerve cells
6 Message passes from nerve cell to nerve cell
7 Message reaches brain
Process ends
8 Brain identifies taste

USE THIS SKILL

Make a Flowchart

Read the passage below about the sense of smell. Make a flowchart that shows this sense process.

The Process of Smell

When you smell a flower, you breathe in chemicals given off by the flower. These chemicals mix with air as they enter your nostrils. The air then passes into the nasal cavity, just below the brain. The chemicals dissolve in the moist lining of the nasal cavity. In the nasal cavity, the hairlike microvilli of receptor cells send messages over nerve fibers to the brain. Finally the brain figures out the messages and tells you how something smells.

TEST TIP

You might be asked to make a flowchart on a test. First shorten each step to a few words. Make a quick sketch of the steps on the flowchart. Then check the sketch against the test passage and make sure your steps are in the right order. Only then should you make a final copy of the flowchart.

Skill 39

HOW TO
Make a Graphic Organizer

Creatures of the Night

What do horseshoe crabs, raccoons, owls, foxes, and scorpions have in common? These very different animals are all far more active at night than they are during the day.

The animals mentioned above are **nocturnal,** meaning that they are awake and active at night and sleep during the day. Just a few of the many North American animals that are nocturnal include alligators, bats, bobcats, crickets, deer, fireflies, grizzly bears, skunks, and wolves. In fact, did you know that most mammal species are nocturnal? Humans and other primates are the exceptions.

To learn more about nocturnal animals, you could collect information by doing research. What if you found so much information that you didn't know what to do with it all? You could use a graphic organizer to help you sort through and organize all this information. A **graphic organizer** is a tool used for organizing information and ideas. Graphic organizers have shapes, such as ovals or boxes, connected by lines. Inside the shapes, you can write key ideas and details. Graphic organizers make information clear, easy to remember, and easy to use. Graphic organizers can also help expand your thinking about a topic. The concept web below is one kind of graphic organizer.

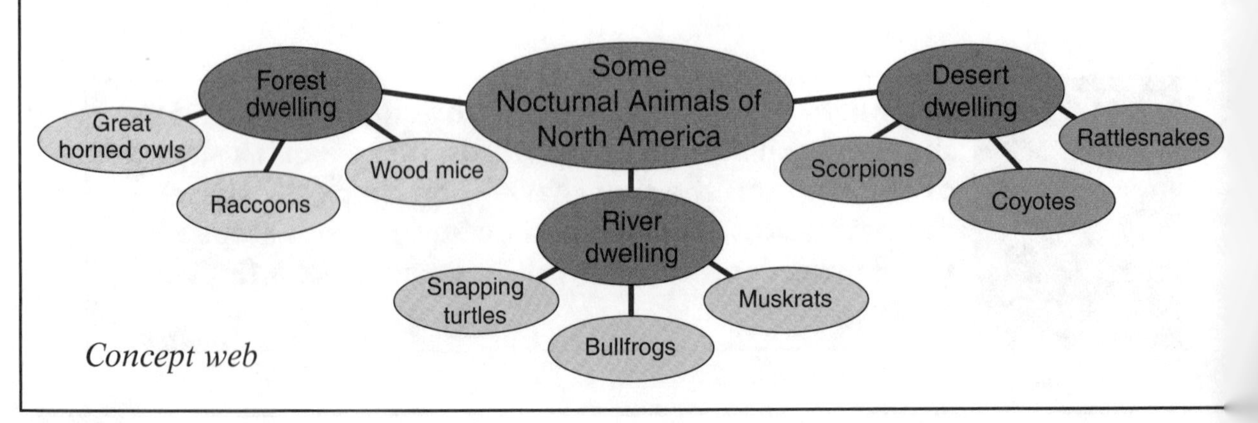

Concept web

STEPS IN Making a Graphic Organizer

Follow these steps to make your own graphic organizer.

1 Decide What to Show

Decide what information you want your graphic organizer to show. Think about how you want to show relationships between facts and ideas. Give your graphic organizer a title that clearly tells what is being shown.

2 Select a Type of Organizer

Choose a graphic organizer that suits the type of information you want to show. These are a few kinds of graphic organizers:

- Use a **spider map** to list the many ideas that come to mind when you brainstorm. You can later connect these ideas with lines to show how they are related.
- Use a **concept web,** such as that on page 194, to organize information on broad topics. List the general topic in the largest oval. Use smaller ovals for subcategories and create still smaller ovals for supporting facts or examples.
- A **network tree** is another type of concept web that begins with main ideas or categories and moves to more specific information. Network trees have connecting information written on or near the lines themselves to explain relationships.

3 Create Your Graphic Organizer

The lines and shapes in your graphic organizer should show relationships among main topics, subtopics, and supporting details. After drawing a rough sketch of your organizer to make sure it suits your topic, you may want to create another version using a ruler, stencils, and any other tools you find helpful. To keep your organizer simple and clear, use words and short phrases instead of longer descriptions.

Network tree

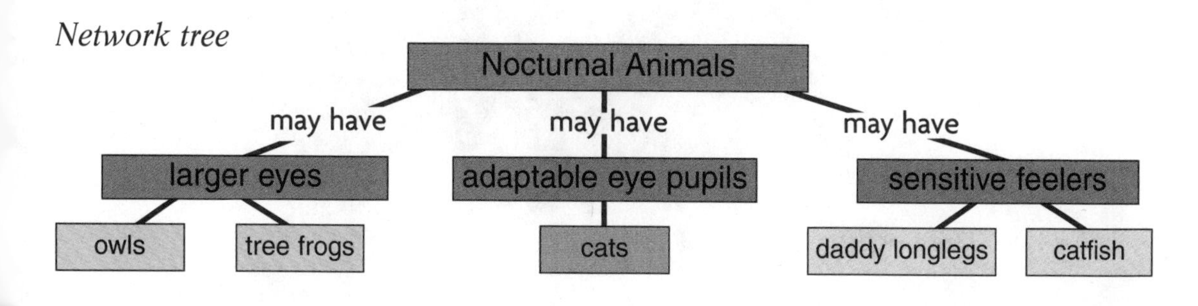

EXAMPLE OF **Making a Graphic Organizer**

See how this spider map shows how certain animals benefit by being nocturnal.

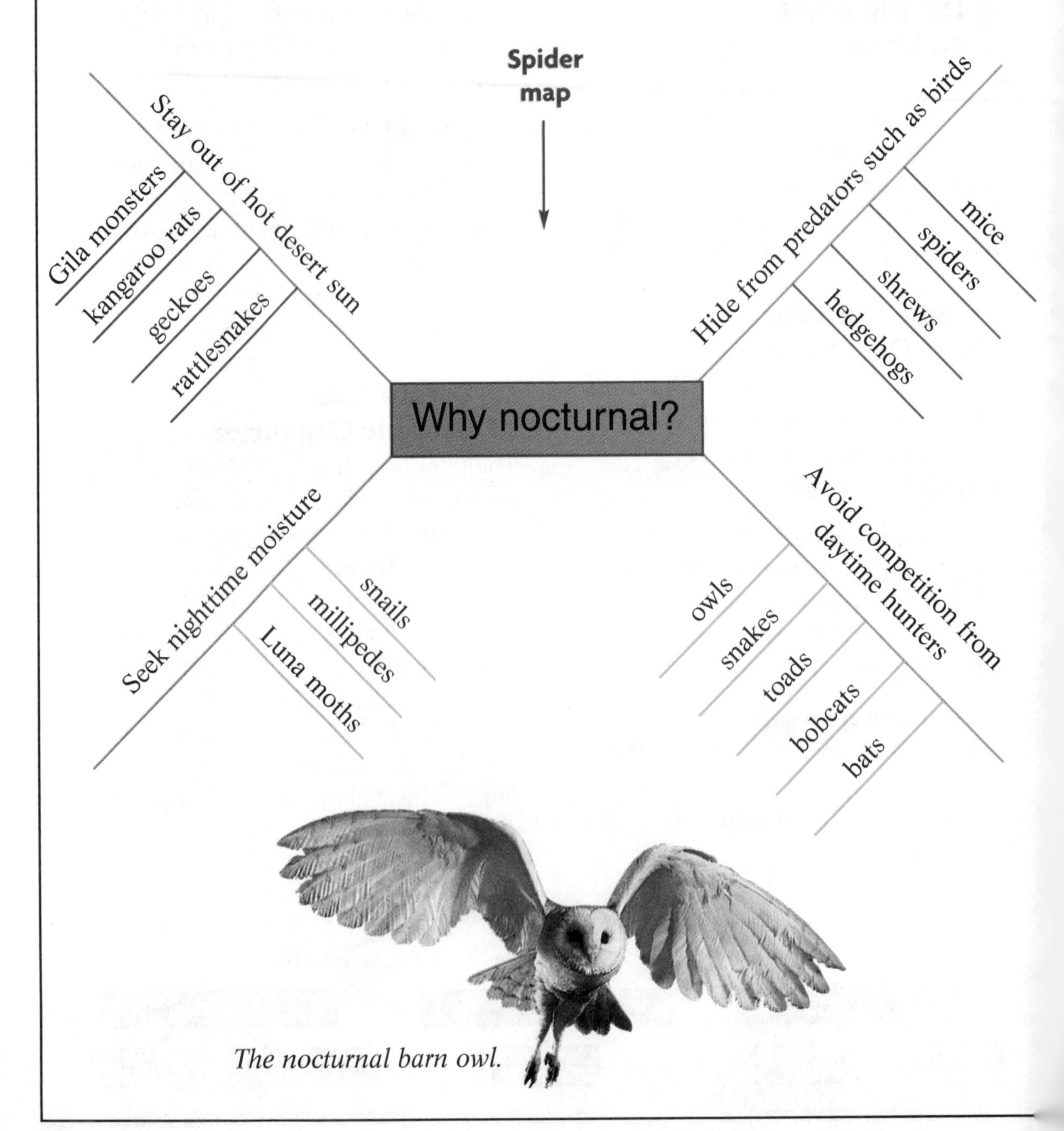

The nocturnal barn owl.

USE THIS SKILL

Make a Graphic Organizer

Read the following passage, then create a graphic organizer to show which nocturnal animals make nighttime sounds, why they make the sounds, and what the sounds are like.

A Symphony of Sound

Have you ever heard a strange noise outdoors at night and wondered what it was? It may have been a nocturnal animal. Nocturnal animals create sounds at night for a variety of reasons. Songbirds call and sing at night to announce that they've found food, that they have claimed a nesting place, or that they are looking for a mate. Male frogs and toads make noises to attract mates. Each species has its own special call that lets females of the same species know where they are. Male insects also make sounds to attract mates, although sometimes their calls are meant to mark territory. When frightened, mice make squeals that are too high-pitched for humans to hear. Rabbits and raccoons sometimes scream when they are frightened or in danger. The raccoon's scream is a loud, "churring" sound, while the rabbit's scream is high and shrill. A mountain lion's loud scream sometimes sounds like a human in trouble. Red foxes and gray foxes make sharp-sounding barks during mating season. Wolves and coyotes howl and "sing" to mark family territory. Wolves may also howl to warn other wolves to stay away or to let pack members know where they are while hunting.

TEST TIP

Graphic organizers are good study tools to use when preparing for tests. Put information into an organizer to summarize it, make it easier to remember, and make relationships clear.

Skill 40

HOW TO

Read a Map

Mapping a Health Risk

When you hear a mosquito buzzing around your head, you probably brush it away or ignore it. In many parts of the world, however, the buzz of a mosquito is reason to worry. Mosquitoes can carry deadly diseases, such as malaria and the West Nile virus.

Malaria is a disease caused by a single-celled parasite. When a female mosquito bites a person who has malaria, it takes in some of the parasites from the person's blood. The parasite then grows for a week or more in the mosquito before it can be passed on. If, after a week, the mosquito bites someone else, the parasites are carried into that person's blood. There the parasite grows and multiplies, eventually making that person very sick.

Malaria is most common in and around tropical areas of the world where mosquitoes thrive. More than 90 percent of all malaria cases are in the area of Africa south of the Sahara Desert. Because of where they live, more than 40 percent of the people in the world are at risk of getting malaria.

Before traveling to other parts of the world, it is wise to find out if malaria or other infectious diseases are common there. One source of information is a **thematic map.** A thematic map shows specific information about different regions. The steps on the next page can help you read a thematic map.

STEPS IN **Reading a Map**

1 Read the Title

The **title** tells the purpose of the map and what information it shows. A map's title usually appears above or below the map.

2 Look for a Legend

The **legend** explains what the symbols mean on the map. The legend is often shown within a box. It is helpful to carefully read the legend before going on to read the rest of a map.

3 Check the Scale and Compass Rose

- A **map scale** compares the distance on the map with the distance in the real world. Different maps use different scales. For example, an inch might equal 100 miles on one map but 1,000 miles on another. Scales usually show distance in both miles and kilometers. Some thematic maps do not use map scales.
- The **compass rose** is a symbol that appears on all maps to show directions. Compass roses show the cardinal directions north (N), south (S), east (E), and west (W). Sometimes only an arrow pointing north (N) is shown on a map.

Worldwide Malaria Risk Areas

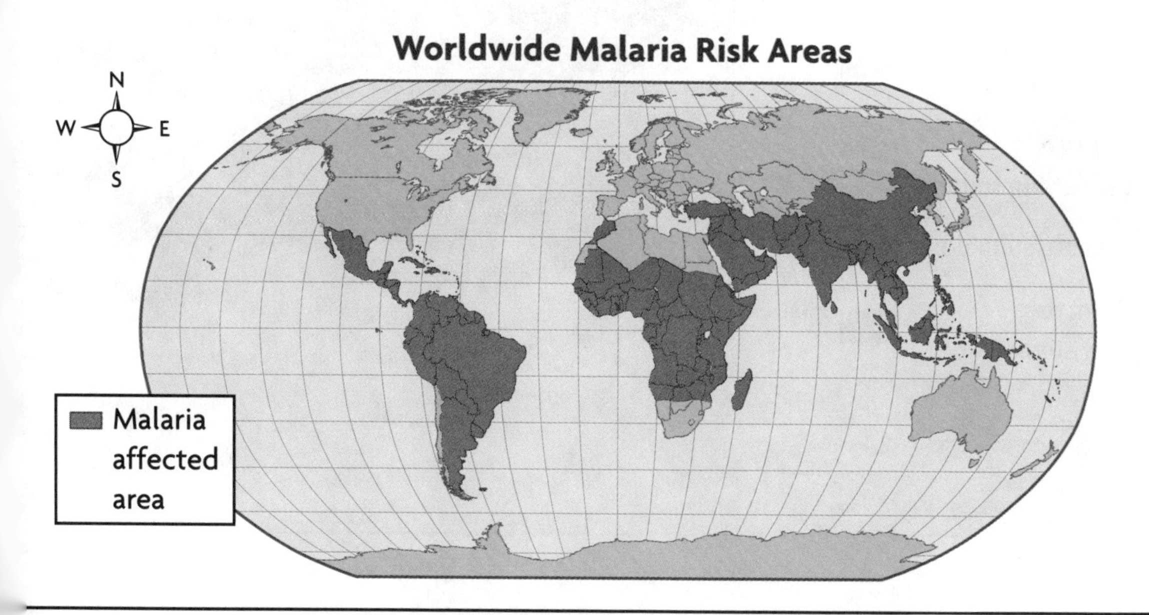

EXAMPLE OF Reading a Map

The map below shows which areas of Africa have climate conditions suitable for the spread of malaria.

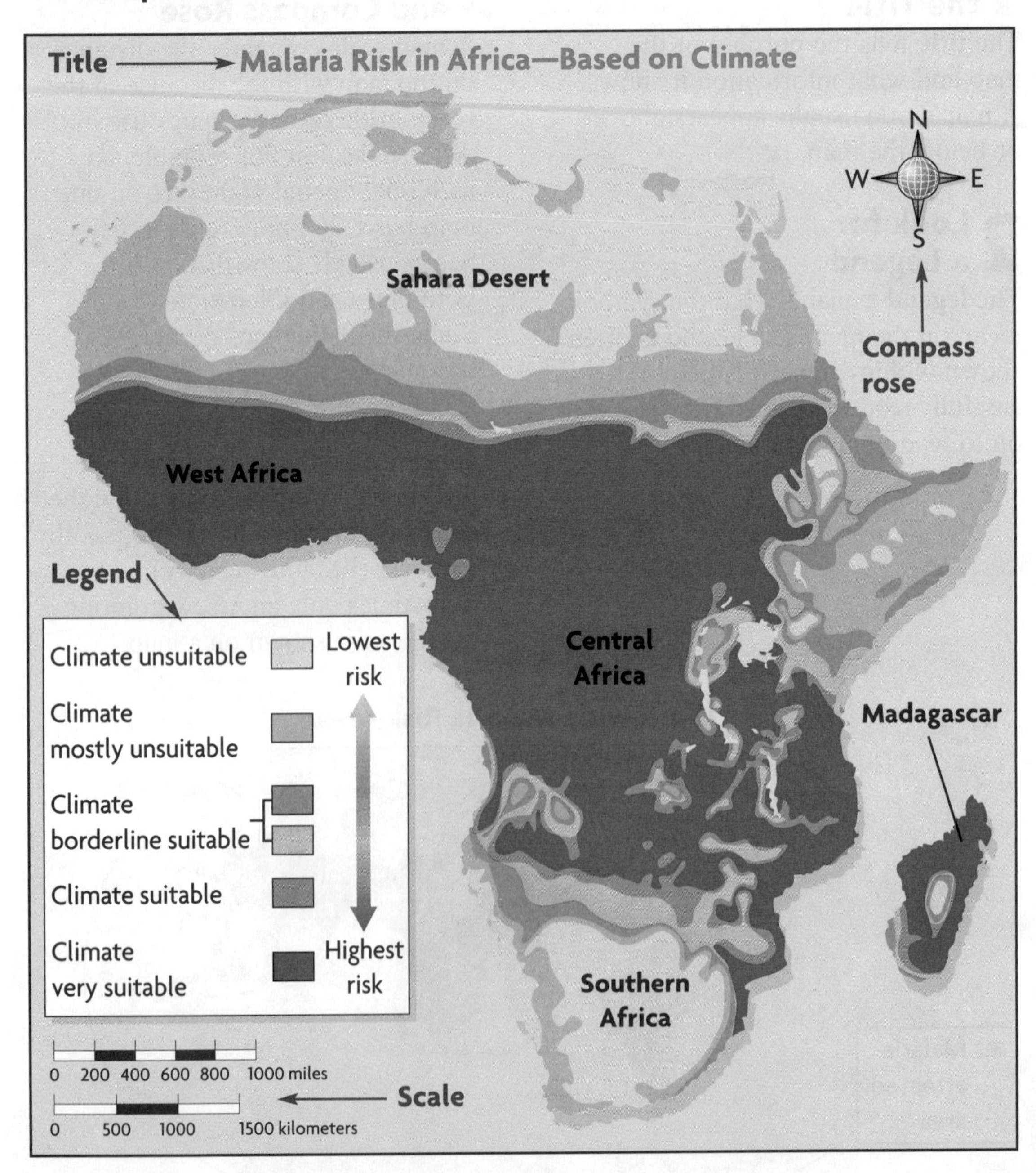

USE THIS SKILL

Read a Map

Refer to the map of Africa on page 200 to answer these questions.

1. What is the distribution of malaria based on in this map?
2. What color represents the area most suitable for the spread of malaria?
3. What color represents the area least suitable for the spread of malaria?
4. The Sahara Desert spreads across the northern third of Africa. According to the map, is this an area where malaria is likely to spread?
5. What other areas of Africa are least likely to have a spread of malaria?
6. Is the island of Madagascar off the east coast of Africa a high risk area?
7. In what part of the continent is malaria most likely to spread?
8. Mosquitoes thrive in rainy, humid, hot climates. What does this tell you about the climate of southern Africa?
9. Which area has a higher risk of malaria, West Africa or northern Africa?
10. Would a traveler be more at risk of being exposed to malaria in central Africa or in southern Africa?

TEST TIP

Before answering test questions about a thematic map, carefully read the map title and look at the meaning of each symbol or color on the legend. Look back at the map to answer each question.

Test-Taking Strategies

Test-Taking Strategy

IDENTIFYING KEY ELEMENTS

On a test some of the information in a passage is more important than other information. The most important pieces of information are called **key elements.** Key elements can tell you about the topic of the passage and help you answer questions. When you read a passage, you don't have to memorize the key elements. Just notice what the key elements are so that you can go back to them when you answer questions.

Read the paragraph below.

Asian elephants and African elephants are very social animals that live in the hot climates of Asia and Africa. Elephants' large ears help them survive in hot temperatures. When elephants flap their ears, air moves across the veins in their ears and cools their blood. This helps keep an elephant from getting too hot. Elephants live in small groups, or herds, made up of female elephants and their young. Living in herds helps elephants survive. Members of a herd warn each other of danger and work together to take care of the young elephants. If one of the members of a herd dies, the other elephants may cover it with twigs and leaves and stay at the grave site for hours. Some male elephants also form herds, but many male elephants live alone. Even though elephants are very strong, some human activities are making it difficult for elephants to survive. Illegal hunting and habitat destruction have caused elephant populations to become dangerously small. The Asian elephant is an endangered species, which means that it is in danger of becoming extinct. The African elephant is a threatened species. A threatened species is one that could become endangered if it is not protected.

Answer the question.

1. Which of the following does **not** affect the survival of elephants?

   A. Members of a herd warning each other if there is danger

   B. Flapping large ears that cool an elephant's blood

   C. Covering a body with sticks and leaves

   D. Illegal hunting and loss of habitat

How to find the answer:

This paragraph contains a lot of information about elephants and their survival. To answer the question, you have to compare the answer choices to key elements in the paragraph. Remember, you are looking for the answer choice that does **not** affect an elephant's survival.

- The paragraph states that "living in herds helps elephants survive." This is a key element of the passage. The paragraph goes on to say that members of a herd warn each other if danger is sensed. This would help an elephant survive, so answer **A** is not the correct answer.
- The paragraph tells that "elephants' large ears help them survive in hot temperatures." This key element shows that answer **B** is not correct.
- The paragraph tells that when an elephant dies "the other elephants may cover it with twigs and leaves and stay at the grave site for hours." Elephants may do this when one elephant dies, but this behavior doesn't seem to have anything to do with the survival of elephants. Answer **C** may be correct. Keep reading to make sure that this is the best answer.
- The paragraph tells that "some human activities are making it difficult for elephants to survive." It also says that "illegal hunting and habitat destruction have caused elephant populations to become dangerously small." These key elements describe activities that affect elephant survival. Answer **D** is not the correct answer choice.
- After comparing each of the answer choices to key elements in the paragraph, answer **C** is the only answer that does **not** seem to affect the survival of elephants. Answer **C** is correct.

Test-Taking Strategy
ORDER AND SEQUENCE

A **sequence** is a list of events that happen over time. In a sequence, some events happen before others. When you find the **order** of events, you tell which events happen first, second, third, and so on. Many tests will ask you questions about a sequence of events in a reading passage or shown on a graph.

Read the paragraph below.

Did you know that rain can sometimes harm fish, destroy buildings, and damage trees and other plants? When the air is polluted, the rain that falls can damage both living and nonliving things in the environment. When cars, factories, and power plants burn fuel, they can release chemicals into the air. Some of these chemicals are acids that mix with water in the air. When this water falls to Earth as rain, it is called acid rain. Acid rain that falls on stone buildings and statues can cause the stone to start crumbling away. Acid rain can damage trees and plants. Acid rain can also pollute groundwater and lakes. Some lakes have been so damaged by acid rain that fish and other water life cannot survive. Due to problems like these, the U.S. Congress passed laws to reduce the pollution that causes acid rain. Many factories and power plants now have devices that filter out some acids before they are released into the air. Each year, there are also new car designs that produce less pollution than earlier car models. Many steps are being taken to reduce pollution and the acid rain that it can cause.

Answer the question.

1. Based on the paragraph, which of the following is in the correct order?

   A. factories release acids, fish in lakes die, acids mix with the water in air

   B. plants are damaged, factories release pollution, devices catch acids

   C. stones crumble, acids combine with the water in air, groundwater is polluted

   D. acids released into air, acids combine with water, acid rain falls to Earth

How to find the answer:

To answer this question, you have to find the answer choice that has the events listed in the same order as the events in the paragraph. The key to answering the question is comparing the answer choices to the information in the paragraph.

- When you compare the events in answer **A** with the events in the paragraph, you know that the last two events in the answer choice are in the wrong order. Answer **A** is not the correct answer.
- By comparing the events in answer **B** with the paragraph, you know that factories release pollution before plants are damaged. So answer **B** is not the correct answer.
- The paragraph tells that stones crumble from acid rain. That means that acids mix with water before stones crumble. Answer **C** is not correct.
- The order of the events in answer **D** matches the order described in the paragraph. Answer **D** is the correct answer.

STRATEGY TIP Sometimes a reading passage will tell about events but will not list the events in order. When you see a passage like this on a test, be sure to look for dates and times in the reading. Dates and times can help you decide which events happened first, second, third, and so on.

Test-Taking Strategy
INFERRING

Most of the time, answers to test questions can be found in the reading passage or on a map or chart. However, sometimes you can't find the answer to the question, and you have to **infer** which answer is correct. Inferring is something you do each day. For example, if you see your brother smiling as he holds his report card, you can infer that he is happy with his grades. When you infer on a test, you use both the information given to you and what you already know to pick the best answer choice.

Read the paragraph below.

It was almost two o'clock on that sunny Thursday in May. People were lining the streets, watching from nearby rooftops, and packed onto boats under the bridge on the East River. It was May 24, 1883, and Abram Hewitt stepped in front of the crowd and announced the opening of the Brooklyn Bridge. Hewitt called the bridge the "crowning glory" of the Industrial Age. From all around, people cheered and celebrated the new bridge that crossed the East River between Brooklyn and New York City. On its opening day, the Brooklyn Bridge was the longest suspension bridge in the world. A suspension bridge is a bridge that is supported by cables or chains between towers at each end of the bridge. When looking across the river in 1883, the towers of the Brooklyn Bridge appeared taller than any other building in New York City except for the thin steeple of the Trinity Church. Unlike many bridges built in the 1800s, the Brooklyn Bridge still stands, thanks to its solid construction and engineering. The two towers that support the bridge do two important jobs. They use huge cables to support the weight of the bridge. They also hold the cables and the roadway high enough so that boats can move along the river beneath the bridge. The Brooklyn Bridge has served as a model for many of the large bridges that have been built throughout the United States.

Answer the question.

1. Which of these is probably true about Abram Hewitt?

   A. He was the mayor of California.

   B. He thought the Brooklyn Bridge was great.

   C. He single-handedly built the Brooklyn Bridge.

   D. He was a captain of a boat on the East River.

How to find the answer:

When you read the question, you can't tell right away that you have to infer to answer the question. When you read the answer choices and compare them to the paragraph, you can see that the answer cannot be found in the paragraph. To find the correct answer choice to this question, you must use information from the passage and what you already know.

- The paragraph does not mention anything about the state of California, nor does it say that Hewitt was a mayor. Answer **A** is probably not correct.
- The paragraph tells that Hewitt called the Brooklyn Bridge the "crowning glory" of the industrial age. You could infer that Hewitt thought that the Brooklyn Bridge was great because he spoke very highly of the bridge in his speech. Answer **B** may be correct, but keep reading to see if there is a better answer choice.
- Because Hewitt announced the opening of the Brooklyn Bridge, you could infer that he might have helped to build the bridge. However, the paragraph states that the Brooklyn Bridge was the longest suspension bridge in the world in 1883. You can infer that the bridge was built by many people and not by Hewitt single-handedly. Answer **C** is not the best answer choice.
- The paragraph tells that Hewitt was on the bridge giving his speech, not on one of the boats. The paragraph does not say anything about Hewitt being the captain of a boat, so answer **D** is not the best answer choice.
- Answer **B** is the best answer choice because it is supported by the information in the paragraph as well as information that you already know.

Test-Taking Strategy
REPHRASING A QUESTION

Some questions on a test can be confusing. A confusing question can make finding the correct answer difficult even if you understand the information on the test. Sometimes you can make a confusing question easier by rephrasing it. When you rephrase a question, you use different words to ask the same question. Just be careful when you rephrase a question and make sure that you include all the important elements of the question.

STRATEGY TIP Be sure not to change the meaning of a question when you use different words to rephrase it.

Read the questions below. Notice how each question is rephrased to make the question less confusing.

1. Of the three types of muscles in your body—cardiac muscle, smooth muscle, and skeletal muscle—which can you control?
   Rephrase: Can you control cardiac, smooth, or skeletal muscles?

2. Which of the following muscles do not belong to the involuntary muscle group, which includes the muscles that line the intestine, the muscles that make up the stomach, and the muscles in the walls of blood vessels?
   Rephrase: Which muscle is not an involuntary muscle?

3. If you compared and contrasted the more than 600 muscles in your body, how would you describe the difference between the muscle that makes up the heart and all other muscles in the human body?
   Rephrase: How is the heart muscle different from other muscles?

Read the paragraph below.

Muscles in the human body can be classified into two groups—voluntary and involuntary. Voluntary muscles are muscles that you are able to control. The muscles in your hands, legs, and face are voluntary muscles. Skeletal muscles are voluntary muscles that allow you to move the bones in your body. Involuntary muscles are muscles that can work all day long without your thinking about them. Your heart and stomach have involuntary muscles. The heart is made up of a type of involuntary muscle called cardiac muscle. Cardiac muscle is found only in your heart. Smooth muscle is another type of involuntary muscle. Smooth muscles are around your stomach, bladder, intestines, and blood vessels.

Answer the question.

1. If you were talking with a friend and shook your head to say "no," which of the following kinds of muscle would you be using to move your head?

   A. Voluntary muscle

   B. Involuntary muscle

   C. Cardiac muscle

   D. Smooth muscle

How to find the answer:

This question may seem difficult and confusing. You could rephrase the question like this: *Which kind of muscle lets you move your head?* There are bones in your head, so you know that you are looking for the kind of muscle that you can control and that allows you to move bones.

- Answer **A** is voluntary muscle, which is the kind of muscle you can control—including muscles that move bones. Answer **A** is correct.
- Answer **B** is involuntary muscle, which is the kind of muscle you cannot control. Answer **B** is not correct.
- Answer **C** is cardiac muscle. The paragraph says that cardiac muscle is an involuntary muscle and is only found in the heart. Answer **C** is not correct.
- Answer **D** is smooth muscle. Smooth muscle is a type of involuntary muscle so answer **D** is not correct.

Test-Taking Strategy
PROCESS OF ELIMINATION

One of the best ways to do well on a test is by using the process of elimination. The **process of elimination** describes a way of finding the right answer by first crossing out answers that are obviously wrong. For example, imagine that you have four possible answer choices to a question. If you can cross out two of the choices that you know are wrong, then you have only two answer choices left. This gives you a much better chance of picking the correct answer.

Read the paragraph below.

Have you ever played music on a CD player? If you have, then you've used a laser! Today lasers are everywhere. In the grocery store, they scan the bar codes on food packages. On streets, the police use lasers to see how fast cars are traveling. At doctors' offices, lasers are making surgeries easier, faster, and safer to perform. In the home, lasers are essential parts of the devices used to play music, watch movies, and even carry phone conversations across fiber-optic cables in phone lines. It's hard to imagine life without lasers, but lasers were invented less than 50 years ago. Light from a laser is a high-energy beam of light that can travel in a single direction. Laser light has wavelengths that are all the same and that travel in step with each other. This makes laser light strong, precise, and very useful in many of today's technologies.

Answer the question.

1. Which of the following is **not** true about lasers?

   A. Lasers are used to see how fast cars are traveling.

   B. Lasers are used to measure the distance to the moon.

   C. Lasers produce high-energy beams of light.

   D. Lasers were invented about 150 years ago.

How to find the answer:

To find the correct answer, compare each answer choice with the paragraph. Eliminate, or cross out, the answers that you know are true statements about lasers. Remember, the correct answer to this question is **not** true about lasers.

- The paragraph tells that police use lasers to see how fast cars are traveling. Answer **A** is a true statement about lasers so it is not the correct answer. Cross out answer choice **A.**
- The paragraph does not mention anything about lasers being used to measure the distance to the moon. You don't know if answer **B** is a true statement or a false statement. Skip this answer choice and keep reading.
- The paragraph tells that lasers are high-energy beams of light. You know this is a true statement so it is not the correct answer choice. Cross out answer choice **C**.
- The paragraph says that lasers were invented less than 50 years ago. Answer choice **D** says lasers were invented about 150 years ago. Answer **D** is a false statement about lasers.
- By crossing out answer choices **A** and **C**, you only have to choose between answer choices **B** and **D.** You know that answer **D** is false but you don't know about answer **B.** Whenever you have to choose between an answer choice that you know is correct and one that you are unsure of, always choose the answer you know is correct. Answer **D** is the correct answer choice to the question.

Test-Taking Strategy

USING CHARTS

Tests often include questions about charts, graphs, maps, and other kinds of graphics that display information. Charts can contain large amounts of information. It is not necessary to read or memorize all the information on test charts. When answering a chart question on a test, read the question first. Then find the information you need to answer the question. Don't be distracted by the extra information on the chart.

Look at the chart below.

This chart shows information about different kinds of stars. The sun in our solar system is a yellow star.

Main Sequence Stars

COLOR	TEMPERATURE	TYPE
Blue-white	About 20,000 °C	B type
White	About 10,000 °C	A type
Yellow-white	About 7,500 °C	F type
Yellow	About 6,000 °C	G type
Orange	About 5,000 °C	K type
Red	About 3,000 °C	M type

Answer the question.

1. Which of the following describes a yellow-white star?

   A. B-type star with a temperature of 30,000 °C

   B. Hotter than a K-type star

   C. A G-type star

   D. Cooler than an M-type star

How to find the answer:

To answer this question, you have to compare each answer choice with the information in the chart. The correct answer choice will have information that tells about a yellow-white star.

- Answer **A** describes a B-type star. When you look on the graph for this information, you can see that this information describes a blue-white star. Answer **A** is not correct.
- Answer **B** describes a star that is hotter than a K-type star. Find the K-type star in the graph. By looking across the row, you can see that a K-type star has a temperature of about 5,000 °C. The chart shows that a yellow-white star has a temperature of about 7,500 °C. A yellow-white star is hotter than a K-type star. Answer **B** does describe a yellow-white star, so answer B may be correct.
- Answer **C** describes a G-type star. The chart shows that a G-type star is a yellow star. Answer **C** does not describe a yellow-white star, so it is not correct.
- Answer **D** describes a star that is cooler than a M-type star. The temperature of a M-type star is 3,000 °C. There are no stars on the chart that have a temperature lower than 3,000 °C. Answer **D** is not correct.
- After comparing each of the answer choices to the chart, you can tell that answer **B** is the correct choice because it is the only answer choice that describes a yellow-white star.

Test-Taking Strategy
USING GRAPHS

Many tests ask questions about graphs. Graphs often show information using numbers, lines, and colors. When you answer a question about a graph, be careful not to make mistakes. Many graph questions will require you to do math or read across long lines. Make sure that you take your time so that you don't make a math mistake or misread information on the graph.

Look at the graph below.

This graph shows information about sounds that some animals make. Sounds with frequencies higher than humans can hear are called ultrasounds. Sounds below what humans can hear are called infrasounds.

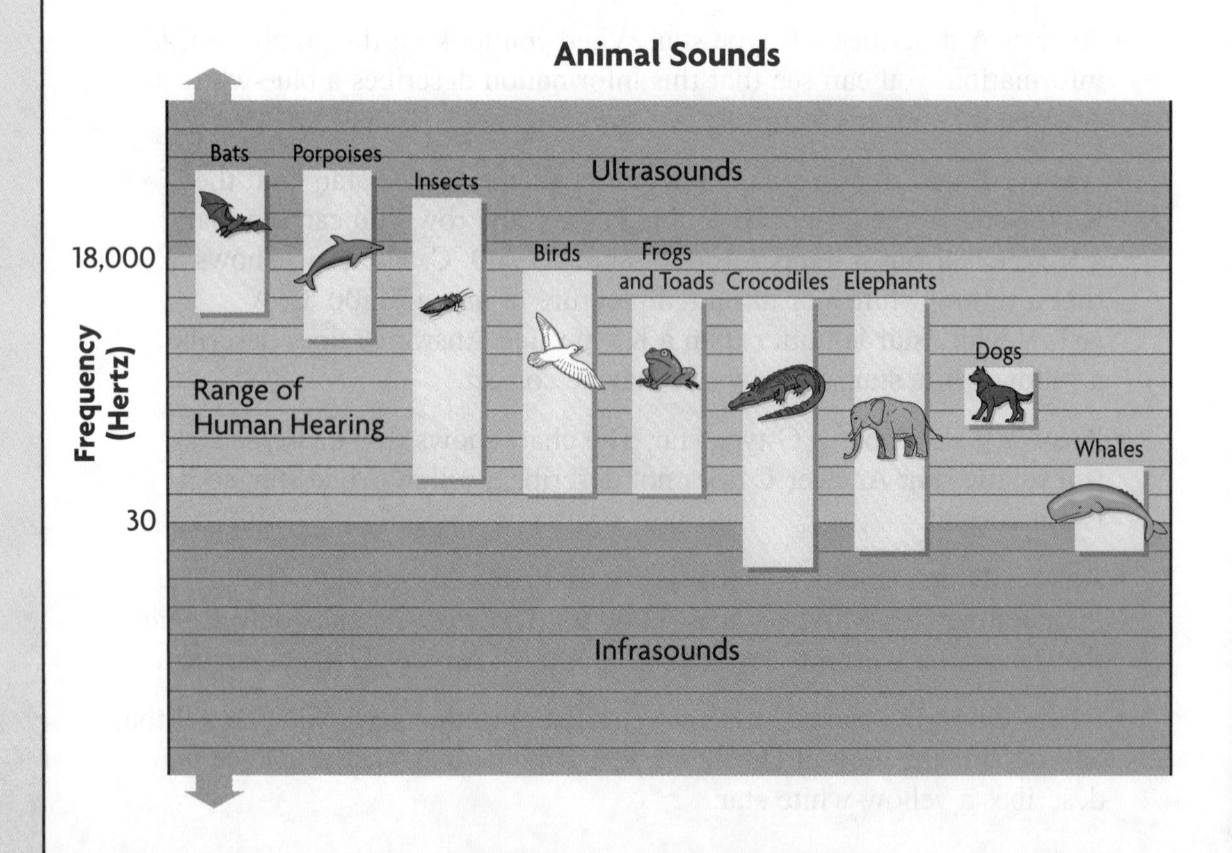

Use the graph to answer the question.

1. Which animals produce sounds lower than humans can hear?

A. Frogs and toads
B. Bats and insects
C. Dogs and porpoises
D. Elephants and whales

How to find the answer:

This graph uses bars to stand for the frequencies of sound made by some animals. The question asks you to tell which animals produce sounds lower than humans can hear. Find the area on the graph that shows the sounds humans can hear. Look at the area below this. This lower area is labeled *infrasounds.* The bars that dip into the area labeled *infrasounds* show animals that produce sounds lower than humans can hear. Compare each of the answer choices with these animals shown in the infrasound area of the graph.

- Answer **A** lists frogs and toads. The graph shows that frogs and toads produce sounds that humans can hear but not sounds lower than humans can hear. Answer **A** is not correct.
- Answer **B** lists bats and insects. On the graph, neither bats nor insects are shown to produce sounds lower than humans can hear. Answer **B** is not correct.
- Answer **C** lists dogs and porpoises. Dogs and porpoises do not have bars that are shown in the infrasound area of the chart. Answer **C** is not correct.
- Answer **D** lists elephants and whales. The bar for elephants and the bar for whales both extend into the area on the graph below the range of human hearing. Answer **D** is the correct answer choice.

A.
B.
C.

STRATEGY TIP Test questions about graphs almost never require you to read the whole graph. Use your time wisely by reading the question first. Then look at the graph's title and the labels. This will help you find only the information you need to answer the question.

Test-Taking Strategy
DEDUCTION

When you make a deduction, you come up with an idea by using a logical pattern of thoughts. Here's an example of a deduction:

On their birthdays, people often blow out candles on a decorated cake. Tanya is blowing out candles on a decorated cake. It is probably Tanya's birthday.

A deduction question on a test usually asks how two things are related or linked to each other. In the example above, blowing out candles is linked to Tanya's birthday. Not all deductions are exactly like the one above, but deductions are always made using a logical pattern of thoughts.

Read the paragraph below.

When Sam was eight, he found out he had allergies. An allergy is the body's reaction to a common substance that the body treats as a serious threat. For example, peanuts are a common food for many people. However, some people's bodies mistake the peanuts for harmful invaders, such as bacteria or viruses, and have allergic reactions to peanuts. When Sam would go outside at the beginning of each summer, his eyes would get itchy and he would sometimes have difficulty breathing. When he went to the doctor about his problem, the doctor did some tests to see if Sam's skin would react to certain allergens. Allergens are substances that can cause an allergic reaction. If Sam had an allergy to a particular substance, his skin would get red where that allergen had been rubbed on or injected into his skin. The tests showed that Sam was allergic to pollen, grass, and dust.

Answer the question.

1. Based on the paragraph, which of the following is probably true?

   A. Sam was not allergic to pollen in the winter.

   B. Sam's skin turned red where pollen, grass, and dust allergens had been rubbed or injected.

   C. Sam was allergic to peanuts.

   D. Sam was not allergic to pollen before the age of eight.

How to find the answer:
To answer the question correctly, you must use information in the paragraph to make a deduction.

- The last sentence of the paragraph tells that Sam was allergic to pollen, grass, and dust. Although the paragraph discusses Sam's allergic reaction in the summertime, there is nothing in the paragraph to make you think that Sam isn't allergic to pollen in the winter. Answer **A** is not correct.

- The paragraph states that if Sam had an allergy to a particular substance, his skin would get red where that allergen had been rubbed on or injected into his skin. The paragraph also says that Sam was allergic to pollen, grass, and dust. Therefore, Sam's skin must have turned red where pollen, grass, and dust allergens had been rubbed or injected. Answer **B** is a deduction that logically links a red skin reaction to allergies. Answer **B** is probably correct, but keep reading just to be sure.

- The paragraph uses peanuts as an example to describe an allergic reaction. The paragraph does not say anything about Sam being allergic to peanuts. Answer **C** is not correct.

- The paragraph starts by saying that Sam found out that he had allergies at the age of eight. The paragraph also describes allergic reactions that Sam had at the beginning of each summer. Because Sam had these reactions each summer, it seems like they had been going on for many years. Answer **D** is probably not the correct answer.

- Answer **B** is the correct answer. It is based on a deduction that can be made using information in the paragraph.

GLOSSARY

A

acid a compound that releases hydrogen ions when dissolved in water. Tomatoes, lemons, and vinegar contain weak acids, whereas car batteries contain much stronger acids.

acid rain rain that is very acidic. Rain can become acidic as a result of pollutants rising from the surface of Earth into the atmosphere.

aerobic exercise rhythmic exercise that increases the heart rate and strengthens the heart and lungs. Running, cycling, and swimming are all forms of aerobic exercise.

allergic reaction when a person's immune system tries to defend the body from a substance that is harmless to most people. These immune responses, instead of protecting the body, can cause sickness and irritation.

almanacs books or booklets that are published yearly and give up-to-date statistics and other information on a variety of different subjects

archeoastronomy the study of how ancient cultures used the science of astronomy in their daily lives

architects professional people who plan, design, and direct the building of structures

astronomy the study of objects in space, such as the moon, stars, and planets

atlases books of maps, used to find out about places in the world

B

bar graph a type of graphic organizer that uses bars, lines, and numbers to show a comparison between two or more things

bob the part of a pendulum that hangs from a string or chain, such as a metal ball or other weight

C

cause the reason something happens

central nervous system the body system that allows sense responses to occur. It includes the brain and spinal cord.

circle graph a type of graphic organizer used to show how parts of a whole are divided up. Sometimes called a pie graph or pie chart.

classify to sort items into groups based on characteristics they have in common

compare to examine qualities that are alike among two or more things

comparison/contrast a written work describing how two things are alike (comparison) and different (contrast)

compass rose a symbol that shows directions on maps

conclusion an explanation of the results of an experiment. A conclusion will tell whether or not data collected during an experiment support the hypothesis.

contrast to examine qualities that are different between two or more things

crystal a solid in which the atoms form repeating geometric patterns. Ordinary table salt is a crystal.

crystallization the formation process of crystals. Some types of mineral crystals are formed when hot, melted rocks cool and solidify.

D

data information such as amounts, measurements, facts, and figures

diagram a sketch, picture, plan, or model used to help explain or illustrate information

E

earthquake the vibrations, or shock waves, caused by movements within Earth

effect something that happens as the result of a cause

encyclopedia alphabetically-organized book of short articles, containing general information about places, people, things, and events

endangered when an animal or plant species becomes so rare, or few in number, that it is in danger of becoming extinct

equilibrium the state in which forces acting upon an object are exactly balanced

era the largest division of time on the geological time scale

extinct when an animal or plant species disappears entirely from Earth

F

fact a statement that can be proven to be true

flowchart a type of graphic organizer that shows all the steps of a process in order

G

generalization a broad statement that is true about *most* members of a particular group

genetic engineering one way that people have tried to improve food. Genetic engineering involves going into the genes of animals and plants and transferring genetic material.

genetically altered a plant or animal in which genetic instructions have been changed

geological time scale the division of the history of Earth into four major time periods based on the kinds of life forms on Earth, as well as the physical changes to the surface of Earth

glossary text that gives definitions of important words found in a book. Provided in the back pages of many nonfiction books, glossaries list words alphabetically.

graphic organizer a tool used for organizing information and ideas. Graphic organizers have shapes, such as ovals or boxes, connected by lines. Key ideas and details are written within these shapes. Some examples of graphic organizers are flowcharts, spider maps, network trees, and concept webs.

guide words words that appear at the top of each page of a reference source as indicators of what articles appear on that page. Sources such as encyclopedias and biographical dictionaries use guide words.

H

horizontal axis the line that runs along the bottom of a graph

hypothesis a proposed answer to a problem or question. Hypotheses are made in statement form, usually as "If . . . then . . ." sentences. An example of a hypothesis is: *"If a pan of cold water is set in direct sunlight for two hours, then the temperature of the water will rise."*

I

index a list of topics discussed in a source, listed alphabetically by keywords or by a person's last name. Many reference sources have indexes on their last few pages.

infer using given information and what you know to answer a question

Internet a computer network that can connect your computer to electronic resources all over the world

involuntary muscles muscles of the body, such as the heart, that work automatically without us thinking about controlling them

J

K

key elements important information in written passages

keywords main words related to the subject of a search or investigation

L

landfill a place set aside for the dumping of garbage

laser an intense, powerful beam of light waves having the same wavelength. *Laser* stands for "**l**ight **a**mplification by **s**timulated **e**mission of **r**adiation."

learning log a notebook or booklet that contains blank pages for recording information about something you are studying

legend map feature that tells what symbols mean

line graph a graphic tool that uses grids, lines, and dots to show how amounts of something change over time

location words words such as *above, below,* and *beneath* that help guide the reader through a description or other written passage

M

magnitude the strength of an earthquake

main idea the topic or focus of a paragraph

making decisions looking at different options and making choices that will help a person or a group reach a goal

malaria a disease caused by a single-celled parasite and spread by mosquitoes. Malaria is a serious health risk in many areas of the world with hot, humid climates.

map scale map feature that compares distances on a map to distances in the real world

mass the term used to tell the amount of matter in an object. Mass is not affected by the force of gravity.

meteorologist a person who studies and predicts the weather. Meteorologists use different kinds of maps and other models in their work.

model a structure or a picture that shows an object, event, or idea. Models can help explain concepts that are hard to see or difficult to understand.

N

nocturnal nocturnal animals are awake and active at night and sleep during the day

O

observation report a written record of what you have seen during a science investigation, written at the end of an investigation. Observation reports give information about all the steps involved in an investigation.

opinion what someone thinks, believes, or feels about something. Opinions are neither true nor false.

oral report one kind of informative report, for which a person gives information about a subject by making a presentation to a live audience

order tells which events happen first, second, third, and so on

outline a structured way to organize ideas or notes. In an outline, information is organized from general ideas to specific ideas and details.

P

pendulum a string or chain attached to a solid object such as a metal disk, then suspended from a fixed point and allowed to swing freely

period the time it takes a pendulum to swing over and back one time. One complete swing, or cycle, in a pendulum's movement.

periods the shorter amounts of time within the four eras of the geological time scale

persuasive writing writing that uses opinions and evidence to try to convince people to think a certain way or to take a certain action

picture diagram a drawing or photo of a subject, with labels that tell about what is being shown

predator an animal that hunts and eats other animals

predicting using past experiences or knowledge combined with present observations to tell what might happen in the future

prey an animal that is hunted and eaten by other animals

problem and solution writing one type of persuasive writing. The author gives details about a problem, then suggests solutions to correct it.

process the steps taken to make or do something

process of elimination finding the correct answer by rejecting all other answer choices

Q

R

recycle to change or process garbage into usable products

reference sources materials that give information about different subjects, such as encyclopedias, almanacs, atlases, dictionaries, nonfiction books, journals, and reports

repeated trials repetitions of a test, done to make sure the results are as accurate as possible

reports written or spoken essays that give information and explanations about events, facts, and ideas

reptiles cold-blooded animals that have backbones, such as snakes, lizards, and turtles

research report a report for which the writer has used reference sources to gather information

Richter scale the scale used to rank and compare the magnitude of earthquakes

S

search engines are computer programs that help you search the Web, or World Wide Web, for information

seismograph an instrument that measures the strength of earthquakes

sequence the order in which events happen

solar energy energy that comes to Earth from the sun

space shuttle a reusable spacecraft

statics a type of science that deals with maintaining equilibrium

summary a short passage that tells the most important ideas from a piece of writing

supporting details facts or examples that give more information about a main idea

survey a set of questions asked of a group of people as a way to gather facts or opinions. Surveys can be spoken or written on paper.

suspension bridge a type of bridge in which the roadway is suspended, or hangs, from steel cables. Suspension bridges are often built over waterways such as rivers and bays.

T

table a type of graphic organizer used to sort information into rows and columns

taking notes writing down key ideas when reading, watching, listening, or thinking

thematic map a type of map that shows specific information about different areas

time line a type of diagram used to list events that happened over a period of time, such as a day, a year, a historical period, or millions of years

title (map) tells what information a map shows

topic sentence often the first sentence in a paragraph, the topic sentence introduces the main idea of a paragraph or other written passage

tropical rain forests forests in areas that get huge amounts of rain every year and have daytime temperatures that stay very close to 80 °F

tropism a plant's response to a stimulus such as light, water, or gravity

U

V

variables the factors that could affect the results of an experiment. For example, variables that could affect the results of a seed-growing experiment are the type of seeds used, the type of soil and fertilizer used, the amount of water given, the temperature of the environment, the growing containers used, and so on.

vertical axis the line that makes up the side of a graph

voluntary muscles muscles of the body that can be controlled by thinking about them, such as the arm muscles

W

weight the term used to tell the specific amount of the force of gravity acting on an object

X

Y

Z

INDEX

U

V

W

X

Y

Z

Credits

Cover, Art Wolf/Stone/Getty Images, Inc., (background) Ralph A. Clevenger/ Corbis; **4** ©Bob Daemmrich/Stock Boston; **6** ©David Young-Wolff/PhotoEdit; **10** ©Corbis; **14, 15** Matt Meadows; **22** ©David Young-Wolff/PhotoEdit; **26** ©David R. Frazier Photolibrary, Inc.; **28, 30** Matt Meadows; **32** ©Carson Baldwin, Jr./Earth Scenes; **33** ©OSF/R. Packwood/Earth Scenes; **35** Matt Meadows; **41** Aaron Haupt; **42** Mark A. Scheider/Photo Researchers, Inc.; **43** Ken M. Johns/Photo Researchers, Inc.; **44** ©Billy E. Barnes/PhotoEdit; **50** Jerry Lodriguss/Photo Researchers, Inc.; **54** Mary Kate Denny/PhotoEdit; **62** ©E.R. Degginger/Earth Scenes; **64** ©John Deeks/Photo Researchers, Inc.; **66** Shambroom/Photo Researchers, Inc.; **70** ©H. Taylor/Oxford Scientific Films/ Earth Scenes; **77** ©Tom & Pat Leeson/Photo Researchers, Inc.; **78** ©Lynn Stone/ Animals Animals; **80** Stone/Getty Images, Inc.; **83** Adam G. Sylvester/Photo Researchers, Inc.; **86** ©Michael P. Gadomski/Photo Researchers, Inc.; **88** ©ASAP/ Photo Researchers, Inc.; **90** ©Craig Lovell/Corbis; **96** ©Phil Dotson/Photo Researchers, Inc.; **104** Matt Meadows; **106** John Mead/Science Photo Library/ Photo Researchers, Inc.; **111** ©Juan Manuel Renjifo/Earth Scenes; **112** ©David Lazenby/Earth Scenes; **115** ©Peter Weimann/Earth Scenes; **116** (l) ©Scott Camazine/Photo Researchers, Inc., (r) ©Wolfgang Kaehler/Corbis; **120** Michael Newman/PhotoEdit; **122** Lawrence Migdale/Photo Researchers, Inc.; **124** Gregory Ochocki; **130** Aaron Haupt; **134** David Young-Wolff/PhotoEdit; **136** Gail Mooney/Corbis; **142** Amy Etra/PhotoEdit; **148** NASA; **151** Matt Meadows; **154** David Young-Wolff/PhotoEdit; **174** ©Tom & Pat Leson/Photo Reseachers, Inc.; **180** David Young-Wolff/PhotoEdit; **196** ©Stephen Dalton/ Animals Animals; **198** ©Jack Clark/Animals Animals.